FARMED AQUATIC FOOD FOR ALL TASTES

The journey of twelve Mediterranean
and Black Sea species
from farms to your plates

Food and Agriculture Organization of the United Nations
Rome, 2023

Required citation:
FAO. 2023. *Farmed aquatic food for all tastes – The journey of twelve Mediterranean and Black Sea species from farms to your plates.* Rome.
https://doi.org/10.4060/cc5140en

ISBN 978-92-5-137777-2
© FAO, 2023

Cover: Ali Elly (illustration) and Chiara Caproni (design).

Contents

Foreword

Farmed aquatic food for all tastes: the journey of twelve Mediterranean and Black Sea species from farms to your plates is the first publication of its kind from the General Fisheries Commission for the Mediterranean (GFCM) of the Food and Agriculture Organization of the United Nations (FAO). It tells the story of twelve farmed species known for their importance or potential in the region, highlighting their history, best farming practices, nutrition and culinary applications along the way.

Given the significance of aquatic food production to food security, employment and economic development and the potential for aquaculture to maintain or increase food production levels while ensuring wild stocks are fished within their maximum sustainable yield, the GFCM is working to sustainably develop this sector through a Blue Transformation, as reflected in the GFCM 2030 Strategy for sustainable fisheries and aquaculture in the Mediterranean and the Black Sea. Achieving this goal means also enhancing the social acceptability of the sector in order to improve its competitiveness and resilience. It was with these aims in mind that this publication was developed.

The preparation of this publication coincided with the International Year of Artisanal Fisheries and Aquaculture (IYAFA 2022) and strives to recognize and empower small-scale artisanal fishers, farmers and workers in these sectors. Through the success stories presented in these pages, the GFCM highlights the valuable contributions of small-scale farmers to building sustainable food systems and promoting food security.

By showcasing the work, efforts and dedication put into producing healthy, sustainable and affordable aquatic food for all budgets and positioning young chefs as ambassadors for these aquaculture products, the GFCM aims to break off the beaten path and open up to the consumers of today and tomorrow.

Miguel Bernal
Executive Secretary
General Fisheries Commission for the Mediterranean

Acknowledgements

Editorial coordinator:
Dominique Bourdenet

Technical coordinator:
Houssam Hamza

Culinary leader:
Julien Ferretti

Editors:
Alexandria Schutte, Matthew Kleiner

Art direction and graphics:
Chiara Caproni

Illustrations:
Ali Elly

Photography:
Nicolas Villion, Daniel Gillet

Technical contributors:
Paolo Carpentieri, Fabrizio Caruso,
Mohamed Chalghaf, Marion Estève, Linda
Fourdain, Maissa Gharbi, Lisa Mionnet,
Elisabetta Betulla Morello, Marie Natalizio,
Christian Née, Georgios Paximadis, Kenza
Tazi, Gautier Wonner

Warm thanks are extended to the Institut
Paul Bocuse Research Centre for its management of
all the culinary components of this publication,
including the development of healthy, tasty and
sustainable recipes, and for its coordination of the
young chefs' participation. The direction, support
and vision of Agnès Giboreau, Research Director of
the Centre, have been fundamental in making this
publication a reality.

We gratefully acknowledge the producers and
young chefs for their precious contributions to the
development of this project. Their stories, insights and
advice have been invaluable.

Producers: Lara Barazi-Geroulanou, Massaad Ejbeh,
Giada Giavari, Jenny Giavari, Joys Giavari, Rodolfo
Giavari, César Gómez, Ercan Küçük, Hasan Kuzuoğlu,
Mohamed Mahmoud Kord, Nadia Selmi, Mohamed
Souei, Brian Takeda, Florent Tarbouriech, Vasyl
Tkachuk, Abu Yazan.

Young chefs: Aurelio Alessi, Lisa Barboteu,
Dilara Cimen, Sarah Fodil, Dor Gali, Nuria Garrido,
Vasileios Konstantinidis, Rihab Naguez, Daniil Nikulin,
Joude El Shennawy, Luca Violi, Chaden Ziadeh.

Special thanks are due to the professionals who
have contributed to the administration, development
and management of this project, in particular
Ysé Bendjeddou, Marion Branchy, Julie Chupin,
Paolo De Donno, Laurence Rispal and Estelle Petit.

Finally, we would like to acknowledge
Intissare Aamri, Lucile Bourdonnec, Adèle Peenaert
and Clara Porcier-Bertels for their contributions to
the communications surrounding this publication.

Farmed aquatic food: tasty, healthy, sustainable

Few foods are as central to Mediterranean and Black Sea culture as aquatic foods. They inspire culinary traditions, provide livelihoods for hundreds of thousands of people and offer a healthy source of protein. From home cooks reaching for handwritten recipes that have been passed down through generations to renowned chefs delicately preparing plates for the most discerning of customers, residents of the Mediterranean and Black Sea region depend on these culinary staples. But how do these molluscs, carp, trout and seabass, to name but a mere few, arrive at their kitchens? It may come as a surprise, but a large portion – almost 3 millon tonnes annually, in fact – originate from a farm.

Across the region, 35 000 farms are engaged in aquatic food farming operations, or aquaculture, as it is commonly called. The sector is active and growing and employs more than 500 000 people either directly or indirectly. It is important as a way to increase aquatic food production without exceeding the natural productivity of wild fish stocks, as a source of jobs and a route for economic development. Enhancing this sector and its benefits are one of the priorities of the General Fisheries Commission for the Mediterranean (GFCM) of the Food and Agriculture Organization of the United Nations (FAO) through its GFCM 2030 Strategy for sustainable fisheries and aquaculture in the Mediterranean and the Black Sea.

However, the benefits of the aquaculture sector are not always obvious to many in the region. It was with this in mind that the GFCM sought to prepare this guide to farmed species showcasing their journey from farm to plate. Within its pages, this guide highlights twelve species chosen for their importance and potential in the region, debunks myths about the sector and emphasizes that farmed aquatic foods can be tasty, healthy and sustainable. Each chapter of this guide focuses on one species, pairing the story of a successful, pioneering producer making waves in the Mediterranean and the Black Sea region with an enticing recipe prepared by the chefs of the Institut Paul Bocuse Research Centre. By the time you reach the back cover, we hope you have a better understanding of the aquaculture sector and are tempted to reach for a farmed species next time you plan to put aquatic foods on the menu.

What is aquaculture?

Before you dive into the recipes and meet the producers, let us give you a better idea of the aquaculture sector in general.

Simply put, aquaculture is the farming of aquatic organisms, be they fish, molluscs, crustaceans or plants. Aquaculture also implies that a human intervenes in the process in some way, for example by feeding the organisms or protecting them from predators.

At the end of the farming process, these organisms can be used in a variety of pharmaceutical, nutritional and biotechnological products, or they can be destined for use as nutritious (and delicious!) foods, for example in one of the twelve tasty recipes you will find in the coming pages. However, farmed species may also be used for another very important purpose: to maintain or increase aquatic food production while reducing pressure on those wild stocks that are being fished above their sustainable levels.

Aquaculture is typically conducted in freshwater, brackish water or in marine environments, depending on the needs of the species. In the success stories presented in this guide, you will read about farmers operating ponds, cages, ropes, tanks and raceways. Pond culture, practised by farming in natural or artificial basins, dates back 4 000 years. On the other hand, cage culture – which allows for farming even in open sea conditions – is a more recent development. Rope culture involves farming by attaching the organisms to a rope submerged in the water and is largely used for oysters, mussels and seaweeds. Tank culture involves rearing species in rectangular or circular units. Finally, raceways are usually systems of multiple rectangular and narrow artificial tanks characterized by continuous water flow. Both tank and raceway culture are often associated with intensive farming practices, allowing producers to achieve higher yields.

Sustainability

There are many misconceptions surrounding the environmental impacts of aquaculture. In fact, aquaculture has the potential to maintain or even increase aquatic food production without increasing pressure on wild stocks and to act as a nature-based solution for recovering stocks and/or ecosystems in bad conditions, and it may even generate a net profit for the local environment. Aquaculture, or more specifically, restorative aquaculture, can do so by helping to regenerate habitats, enhancing and restocking wild stocks, improving the health of ecosystems via carbon sequestration and water quality improvement, and making ecosystems more resilient to climate change. In the coming years, the sustainability of aquaculture will be further strengthened thanks to the launch of the FAO strategy for the Blue Transformation of aquatic food systems, which aims in part to enhance the development of aquaculture while minimizing its environmental impact, as well as the GFCM commitment, through its 2030 Strategy, to secure animal health and welfare.

Many farms in the region, such as those featured in this publication, have made improving their sustainability a priority, including through better feeding practices, switching to renewable sources of energy and enhancing their production methods.

Beyond its potential environmental benefits, aquaculture can also have positive social and economic impacts, thus helping to secure livelihoods and sustainable food systems. These factors have made it a vital sector for the achievement of the 17 United Nations Sustainable Development Goals (SDGs), including Goal 14: Conserve and sustainably use the oceans, seas and marine resources for sustainable development.

The species

This publication features twelve species. They were each chosen for their popularity, history, importance and potential in the region.

- **Common carp** (*Cyprinus carpio*): This fish is one of the most important freshwater species farmed worldwide and is popular as an affordable and nutritious source of protein. Serve it as **carp balls with a smoky eggplant cream** (p. 9).
- **Mediterranean mussel** (*Mytilus galloprovincialis*): Adaptable and able to inhabit a wide variety of environments, it has become one of the world's most cultured molluscs. Taste it for yourself as **mussel skewers with a rosemary and pomegranate sauce** (p. 21).
- **Gilthead seabream** (*Sparus aurata*): Thanks to its high survival rate and feeding habits, it is well-suited to aquaculture and a popular choice

among farmers. Enjoy it at home **grilled and stuffed with green fennel, orange and capers** (p. 33).
- **Flathead grey mullet** (*Mugil cephalus*): This fish has a long history of farming in the region and is known for its hardiness, simple diet and rapid growth. Try it **breaded and pan-fried and served with tarragon aioli** (p. 45).
- **Stony sea urchin** (*Paracentrotus lividus*): This spiny, round echinoderm is an important herbivore in the region and highly appreciated for its unique taste. It makes for a delicious meal when served in a **risotto with chanterelles and tangerine** (p. 57).
- **Seaweed** (*Gracilaria* spp. and *Ulva* spp.): It has been harvested in the region since ancient times, but marine farming is relatively new. Try this aquatic ingredient for yourself in a **tajine with pistachios and dates** (p. 69).
- **Tilapia** (*Oreochromis* spp.): As tropical fish, they thrive in warm waters and their omnivorous behaviour allows for rapid growth. They are excellent served as a Turkish-inspired **sandwich with confit red pepper, grilled zucchini and a dill and tahini dressing** (p. 81).
- **Turbot** (*Scophthalmus maximus*): Prized for its mild flavour and health benefits, this fish has seen rapidly increasing aquaculture production. Enjoy it **pan-fried and paired with an eggplant tian and arugula and basil pesto** (p. 93).
- **European seabass** (*Dicentrarchus labrax*): This predator species was the first non-salmonid marine species to be commercially cultured in Europe and can inhabit waters with a wide range of salinities and temperatures. Try it as a Mediterranean-inspired **caponata-stuffed seabass in a bread crust** (p. 105).
- **Rainbow trout** (*Oncorhynchus mykiss*): A favourite of farmers around the globe since the nineteenth century given its fast growth, easy spawning and tolerance for a range of environments. Try it **baked with sage and served with a zucchini and walnut condiment** (p. 117).
- **Pacific oyster** (*Magallana gigas*): The oyster of choice for producers in many regions of the world, due in part to its rapid growth and resilience against environmental conditions. It is excellent prepared as a **tartare alongside peaches and tomatoes** (p.129).
- **Beluga sturgeon** (*Huso huso*): With roots dating back hundreds of millions of years, it is the oldest

freshwater fish in the world. Due to its biology, wild populations are mostly protected, but its rearing in aquaculture allows for the creation of jobs and the production of caviar and very tasty meat that is wonderfully showcased **baked in a salt crust and served with fregola sarda risotto** (p. 141).

What to look for in aquatic foods

Another great benefit of aquaculture is that it can make traditionally expensive foods more affordable, allowing you to eat quality aquatic foods without emptying your wallet. When you can, it is advisable to give preference to aquatic foods that are minimally processed and to limit the consumption of prepared foods such as breaded fish, fish in sauce, croquettes, rillettes and fish or shellfish mousse, to name a few.

When at the market, keep in mind that one portion of fish is equal to 100 g and it is recommended to eat two of these portions per week, one of which should be an oily fish high in polyunsaturated fatty acids and omega-3 fatty acids.

Consider purchasing aquatic foods from a small-scale aquaculture producer in your area. These producers play an important role in coastal and rural communities by contributing to food security, livelihoods and well-being. By buying products from local small-scale producers, you are helping to support and empower them and to stimulate the economy within your community.

Nutrition

Not only are aquatic foods delicious and versatile ingredients in traditional and modern cuisine, they also have great health benefits. Rich in protein, good fats, vitamins and minerals and low in carbohydrates, these foods are an excellent choice for your table and are a key pillar in the renowned Mediterranean diet.

The average protein content per serving of fish is an impressive 19 g, although it is slightly lower for shellfish. Even better, the proteins are of good quality and are highly digestible, meaning that they will be very well assimilated and used by the body.

Compared to meat, aquatic foods are relatively low in saturated fatty acids, which, if consumed in excess, can have a negative impact on health. However, they are rich in good-quality fats, making them an important food, as these fats play an essential role in the body as part of the composition of our cells, a source of energy and a key factor in hormonal functions. Fish are particularly high in omega-3 polyunsaturated fatty acids. These fatty acids are essential to healthy functioning, help in preventing the onset of cardiovascular disease, contribute to proper brain functioning, have a positive impact on the nervous system and can even help reduce the risk of certain cancers. Omega-3 fatty acids cannot be synthesized by the body and must be provided by the diet; therefore, it is recommended to increase the consumption of foods rich in omega-3 fatty acids, such as aquatic foods.

Fish is also one of the richest animal sources of vitamin B_6, which is involved in many physiological reactions and in protein renewal, as well of vitamin B_{12}, an essential vitamin for the body. Given their high levels of vitamins D, A and E, aquatic foods support bone health, immunity and healthy vision, and help protect against chronic degenerative diseases.

Many essential minerals are found in aquatic foods. Adding more fish and seafood to your diet will help to meet your daily requirements of potassium, phosphorus, calcium, sodium, selenium, iron, iodine and zinc. These minerals play key roles in the body, from helping to support bone health to regulating blood sugar levels to maintaining the body's water balance.

Now that you have discovered more about the ins and outs of aquaculture and the benefits of aquatic foods, it is time for you to embark on the journey from farm to plate. Along the way, we hope you enjoy the stories of producers from throughout the region, uncover the realities of the aquaculture sector in the Mediterranean and Black Sea and are inspired by the twelve tempting recipes to come.

Common carp

Cyprinus carpio

Common carp in the Mediterranean
and Black Sea region

Scientific name:
Cyprinus carpio

Family:
Cyprinidae

Approximate average
production volume (2016–2020):
153 200 tonnes

The top three producers are
**the Russian Federation,
Egypt** and **Ukraine**.

100% of production
occurs in **freshwater**.

An affordable source of
protein that is rich in **vitamins**
and **minerals**.

Source: FAO. 2023. Global aquaculture production quantity (1950–2020).
In: *Fisheries and Aquaculture Division.* Rome. Cited 26 July 2022.
fao.org/fishery/statistics-query/en/aquaculture/aquaculture_quantity
Based on data from the GFCM Information System for
the Promotion of Aquaculture in the Mediterranean (SIPAM)

Cyprinus carpio

Common carp

C arp is one of the most important farmed fish worldwide, as it is easy to farm, resilient and adaptable to many aquaculture techniques. In the Mediterranean and the Black Sea, many species of carp are commercially produced, but common carp (*Cyprinus carpio*), a species of the Cyprinidae family, is by far the most prevalent (FAO, 2023). Known as an affordable and nutritious source of protein, carp has been an important food source for thousands of years and continues to be a staple on many menus today.

Culinary and nutritional value

Considered a semi-fatty fish, carp is a significant source of protein, vitamins and essential minerals. The dense and slightly pink flesh of this affordable, low-calorie fish can be prepared in many ways: roasted, grilled, braised, stuffed or poached in a soup. Mincing the flesh also makes it an ideal ingredient for gratins or fish balls.

Farming in the region

In the Mediterranean and Black Sea region, the production of common carp is concentrated in Egypt, the Russian Federation and Ukraine, which together produced approximately 152 500 tonnes in 2020 (FAO, 2023). Cyprinids have been reared in China for 2 000 years and common carp was among the first freshwater species to be introduced to Europe (Copp *et al.*, 2005). Despite this long history, the first step toward the domestication of common carp did not take place until between the twelfth and mid-fourteenth centuries. In the nineteenth century, European producers began the controlled semi-natural breeding of carp in ponds and the rearing of juvenile

carp (FAO, 2022a). Today the species continues to be of great interest to farmers given its adaptability to changing conditions (Manjappa *et al.*, 2011; Rahman, 2015). The production cycle takes place over three main phases: seed supply, nursing and ongrowing. The most effective method of establishing a seed supply involves inducing spawning in part by maintaining water temperatures of 20–24°C. Following a short stay in large conical tanks, hatched juveniles, also called fry, are moved to shallow, drainable ponds – or tanks if there are many predators – to nurse for three to four weeks. During this nursing period, carp juveniles are fed with rotifers,

a type of microscopic animal, supplemented by soybean meal, cereal meal, meat meal, rice bran or rice polishings. They are then moved to semi-intensive ponds to grow until they reach fingerling size (FAO, 2022a). The subsequent ongrowing stage can last anywhere from 120 to 170 days for carp to reach sizes up to 0.5 kg and from 240 days to 340 days to reach sizes up to 3 kg, after which farmers proceed with harvesting, usually using seine nets (FAO, 2022b).

The rural farmer's enterprise Dzherela: modernization in the interest of food security

The Ukrainian rural farmer's enterprise Dzherela has become a pioneer in the field of selective carp breeding. Based in Radyvyliv, Ukraine, the enterprise works to enhance food security in the region by increasing the productivity of carp farming and improving the species' resistance to various diseases, stress and industrial technologies. In cooperation with the Institute of Fisheries of the National Academy of Agrarian Sciences of Ukraine (NAASU), Dzherela specializes in the cultivation, reproduction, and artificial selection of Ukrainian carp breeds, namely Halych and Lubin carp, and the production of hybrids using broodstock of Amur carp (*Cyprinus rubrofuscus*).

But Dzherela did not always specialize in the field of selective breeding, or even aquaculture for that matter. In 2002, it began as a breeder of Carpathian bees and a cultivator of plants including soybeans, thistles, naked oats and wheat. Bursting with a strong enthusiasm to create a full-fledged system of environmentally friendly, organic and local products, the enterprise quickly expanded into the aquaculture sector, renting and performing a major reconstruction of nearby fishponds, and began to farm a surface area of 40 ha of water by the end of 2002. Inspired by its initial success, the enterprise increased its production capacity through the reconstruction of hydraulic structures, drainage of wetlands and creation of new ponds. Today, the enterprise boasts 500 ha of land, of which 200 ha are used to farm oats, soybeans, milk thistle and wheat crops for carp feed and 300 ha are used for aquaculture.

In 2006, the enterprise began cooperating with the NAASU Institute of Fisheries in the field of selective breeding, focusing on the creation of broodstock of the Lubin intra-breed type of carp. Four years later, it began reproducing carp indigenous to Ukraine – specifically Halych carp, which had previously shown promising performance. Today, the enterprise continues its work in the field of selective breeding and crosses the highest-quality Halych carp females with Lubin carp males to create rapid-growing, resilient fish. Driving this work is the enterprise's mission to improve the production of domestic aquaculture, promote the development of rural areas by organizing modern production, increase employment rates, obtain new lines of carp for industrial cultivation and provide the domestic market with healthy, organic fish products.

Breeding works at the farm start in the early spring when the water temperature reaches 10°C. Senior fish breeders conduct a comprehensive assessment of the fish according to their breeding and productive qualities to determine their grade. Based on this assessment, each fish is assigned an appropriate class – elite, first or second – with elite carp favoured for spawning. After spawning, captured brooders are transferred for fattening until the autumn and again for wintering. The production of commercial fish and fish seeds begins in late autumn with the preparation of nursery and fattening ponds.

©Dzherela

"Carp is a commonly eaten fish in France, along with seabass and trout. The carp is delicious cut into pieces and fried, served with lemon juice or mayonnaise. A light salad brings some freshness to the dish."

Aurelio Alessi, Culinary Student at Institut Paul Bocuse

Following spawning, fry are placed in nursery ponds for cultivation. The second week of October marks the beginning of the harvest.

The enterprise has dedicated over 20 years to supporting local markets and producers and enhancing food security in the region. At the onset of the war in Ukraine in 2022, the enterprise expanded its provision of aquatic products without hesitation and was one of the first in the aquaculture sector to support the state's food security in terms of the provision of aquatic bioresources under martial law.

As it continues to enhance production and works to build a collection of broodstock to produce carp indigenous to Ukraine, the enterprise has identified the need to improve its regulatory environment. Dzherela calls for the process of leasing water bodies for aquaculture purposes to be updated, particularly regarding restrictions on the use of general water for water bodies used to keep breeding and genetic material, as well as for better streamlining of legal frameworks and an increase in the level of state support. Rises in the costs of production (fuel, feed, fertilizer, etc.) coupled with consumers' reduced purchasing power has also led to a situation in which the enterprise faces greater expenses that it cannot recuperate by increasing prices.

In cooperation with the NAASU Institute of Fisheries, the enterprise aims to continue enhancing knowledge around selective breeding by consolidating the status of intrabreed carp species, conducting additional genetic studies of brood fish and creating a database based on genetic passports. Through these actions, Dzherela aims to legitimize the existing gene pool of Ukrainian carp breeds and further promote the development of selection and breeding relationships with European and international fish farms and research institutions. Simultaneously, the enterprise plans to introduce an artificial method of reproduction that involves the construction of an incubation facility with the possibility of growing larvae to viable stages under controlled conditions.

"Our aim is to enhance carp production and food security in the region through cutting-edge work in the field of selective breeding."

Traditions and recipes from the region

Carp has a long tradition in the culinary history of the Mediterranean and the Black Sea. The French gastronomist Jean Anthelme Brillat-Savarin (1755–1826) described carp as a noble fish and cooked it stuffed whole with lard, bread and crayfish. The roe was also used to flavour omelets. Over the last century, carp gained an unlucky reputation as a low-quality fish with a muddy taste. Advances in aquaculture practices, however, have helped to achieve fish with a delicate flavour.

©GFCM/Selvaggia Cognetti de Martiis

Israel / *Gefilte* **fish**

This Ashkenazi recipe of minced carp flesh is either served poached as a meat ball or used as a filling for baked white fishes. The carp are filleted, the skin is removed, and the heads, fins and skin of the fish are used to prepare a fish fumet with onions, carrot, thyme, laurel and black pepper. Once the flesh has then been blended with matza breadcrumbs and flour, hard boiled eggs and salt, the mixture is shaped into balls and poached in the prepared broth until cooked through. The fish balls are traditionally paired with a red sauce made with horseradish and beetroot.

©GFCM/Dominique Bourdenet

Montenegro / *Krap u tavu*

Translated to "carp in a pot", this traditional recipe from the Montenegrin region surrounding Lake Skadar is prepared with scaled and gutted carp. The skin is carefully incised with a sharp knife before the fish is quickly pan-fried on both sides. A sweet and savoury mixture prepared with sauteed onions, apples, quinces, plums, tomatoes, flour and white wine vinegar accompanies the fish. The carp is placed in a pot and topped with the sweet and sour preparation. It is then baked in a preheated oven until the inner flesh is cooked.

©GFCM/Sinziana Demian

Romania / *Ciorbă de peşte*

This traditional Romanian soup is commonly prepared with carp from the Danube River. The fish are either scaled and gutted, kept whole or cut into pieces including the head and fins. The fish is then poached in a slightly acidic broth prepared with a fermented wheat bran called *borş*. Vegetables such as carrots, celery, onions and potatoes are diced and tossed into the soup. Parsley and lovage leaves are added at the end of the cooking process to flavour the broth. Some versions of *ciorbă* also contain tomatoes and red bell peppers.

Chef's tips

How to prepare carp?
The taste of carp is better when consumed fresh. While selecting the fish, pay attention to the brightness of the eyes, the red colour of the gills and the shininess of the skin. To prepare the whole fish at home, use the back of a knife to remove the scales without piercing the skin and use fish tweezers to remove the fishbones. Rinse the tweezers in a bowl of water to prevent the fishbones from sticking.

How to cook carp?
Carp can be prepared in a diversity of ways, from on the grill or an open fire to poached in a soup. When poaching carp, be careful not to bring the soup or broth to a simmer with the fish inside, as it will become dry. If the carp is filleted, carefully remove the cheeks with a small spoon. They are delicious pan-fried with a dash of olive oil and a pinch of salt.

How to season carp?
Though carp is often associated with a muddy taste, the fish from aquaculture farms are grown in fresh and clean water and therefore develop a mild, aromatic profile. As carp flesh lacks an intense flavour, it can be seasoned with a multitude of spices and herbs. To sharpen the flavour of the carp, marinate the fillets with white wine or citrus juice and aromatic herbs.

Here are the best ingredients to pair with carp for unique Mediterranean and Black Sea-inspired dishes

Vegetables
- Potato
- Tomato
- Onion
- Turnip
- Green peas
- Carrot

Fruits
- Lemon
- Orange
- Grapes

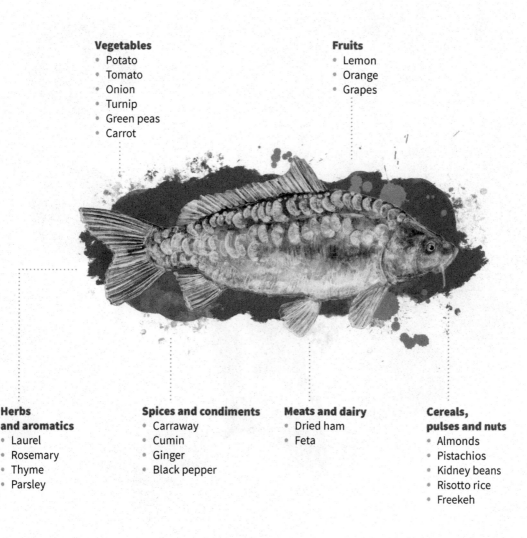

Herbs and aromatics
- Laurel
- Rosemary
- Thyme
- Parsley

Spices and condiments
- Carraway
- Cumin
- Ginger
- Black pepper

Meats and dairy
- Dried ham
- Feta

Cereals, pulses and nuts
- Almonds
- Pistachios
- Kidney beans
- Risotto rice
- Freekeh

Carp balls with smoky eggplant cream

Type
starter
Yield
4 servings
Preparation time
20 minutes
Cooking time
10 minutes

This dish sees carp blended with the oriental flavours of ras el-hanout, coriander and raisins and fried in the same manner as falafel.
The carp balls are served with a smoky eggplant sauce inspired by the Lebanese baba ganoush.

Utensils needed
* Cutting board
* Baking tray
* Greaseproof paper
* Blender
* Mixing bowl
* Spoons
* Deep fryer
* Bowl

Ingredients

Eggplant sauce
4 eggplants
Salt and pepper
1 garlic clove
1 tbsp tahini
Juice of 1 lemon
4 tbsp extra virgin olive oil
1 tsp smoked paprika

Carp balls
4 carp fillets
½ white onion
3 slices of soft bread
1 egg
½ bunch of fresh coriander
Juice of 1 lemon
1 handful of raisins
½ tbsp ras el-hanout
Salt and pepper
Canola oil for frying

Topping
1 pomegranate

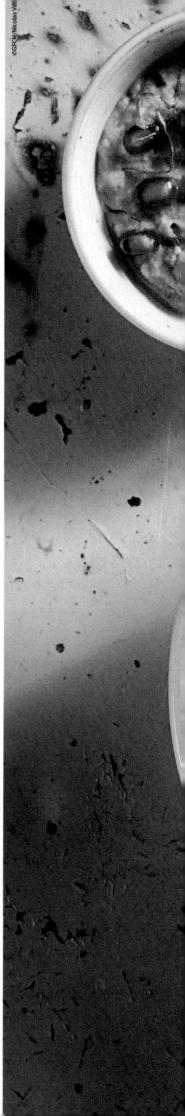

Preparation

Eggplant sauce
Slice the eggplants lengthwise and place on a baking tray lined with greaseproof paper. Sprinkle with olive oil and season with salt and pepper. Bake in an oven preheated to 180 °C for 30 minutes. Remove the eggplant flesh from the skin and blend it with garlic, tahini, lemon juice, smoked paprika and olive oil. Season with salt and pepper.

Carp balls
Remove the fishbones and skin of the carp fillets and rinse them under cold water. Finely chop the onion and blend with the carp fillets, soft bread slices, egg, coriander, lemon juice and raisins. Season with ras el-hanout, salt and pepper. Using two spoons, shape the fish batter into quenelles. Deep-fry the quenelles for 3 minutes at 180 °C, until golden-brown.

To serve
Cut the pomegranates in half and remove the seeds. Place the carp balls in a large bowl accompanied by the eggplant sauce topped with pomegranate seeds.

Nutrition facts

	Per 100 g	Per recipe
Energy	552 kJ/132 kcal	13 359 kJ/3 193 kcal
Protein	7.1 g	172.0 g
Carbohydrate	3.8 g	92.1 g
Fibre	1.6 g	39.2 g
Sugar	2.7 g	64.9 g
Fat	9.4 g	226.0 g
Saturated fat	1.3 g	32.3 g
Sodium	58.6 mg	1 417 mg

Mediterranean
mussel

Mytilus galloprovincialis

Mediterranean mussel in the Mediterranean and Black Sea region

Scientific name:
Mytilus galloprovincialis

Family:
Mytilidae

Approximate average
production volume
(2016–2020):
320 300 tonnes

The top three producers
are **Spain**, **Italy** and **Greece**.

99% of production
occurs in **marine water** and
1% in **brackish water**.

A **healthy**, **quick to cook** and
affordable species that is
a staple on many coastal
restaurant menus.

Source: FAO. 2023. Global aquaculture production quantity (1950–2020). In: *Fisheries and Aquaculture Division*. Rome. Cited 26 July 2022. fao.org/fishery/statistics-query/en/aquaculture/aquaculture_quantity

Based on data from the GFCM Information System for the Promotion of Aquaculture in the Mediterranean (SIPAM).

Mytilus galloprovincialis

Mediterranean mussel

Originating from the Mediterranean Sea, the Mediterranean mussel (*Mytilus galloprovincialis*) has grown to become one of the world's most cultured and commercialized molluscs (Turolla, 2016). In Europe alone, this bivalve accounted for almost 300 000 tonnes of production in 2020 (Eurostat, 2022). This success is due in part to the species' adaptable nature, ability to inhabit both offshore areas and lagoons at a wide range of temperatures and oxygen levels, and the low level of technology needed for its production.

Culinary and nutritional value

Mussels' affordable price and quick cooking process make them a valuable choice for coastal restaurants. Mussels have high concentrations of protein and omega-3 fatty acids and their flesh is full of iodine flavors. They can be consumed raw with lemon juice or as a tartare or even served warm, either steamed, in a soup, deep-fried or grilled on an open fire.

Farming in the region

The earliest evidence of commercial Mediterranean mussel culture in the Mediterranean Sea dates back to 1901 in Tarragona, Spain, where farmers used poles as supports for the mussels' growth. Eight years later, this practice expanded to Barcelona but was abandoned in favour of the use of floating structures. By 1946, raft culture, which involves attaching mussel seeds to ropes hanging from a floating raft, was introduced and production took off (FAO, 2022c). Among these systems, raft culture remains in use today, although producers often opt to suspend ropes from a wooden frame or suspend longlines from floating buoys.

Mussel farming is an extensive aquaculture practice. This means that during the production cycle, no feeds or antibiotics are given and instead, everything the mussel needs to grow is provided by the environment. High-quality waters are therefore of paramount importance. Meanwhile, producers are responsible for providing a solid substrate for the mussel to grow, managing the population and maintaining the farming structures.

To begin the culture process, producers attach mussel seeds to the suspended ropes. Five to six months later, when the mussels have reached half their desired market size, producers begin the thinning process to prevent the mussels from falling off the ropes and to encourage rapid and uniform growth. To do so, producers lift the ropes from the water and remove the mussels by hand. They then sort the mussels by size and reattach them onto new ropes. This process is repeated to ensure that all the mussels reach a similar size by the time of harvest. It can take anywhere from eight to thirteen months for mussels to reach market size, although many producers opt to use ropes to collect seed, grow mussels and retain market size mussels all at once in order to secure year-round production.

Peak harvest time varies depending on environmental conditions, but usually occurs just prior to the reproductive season, when the mussels' nutritional quality is at its best. Producers use cranes to lift the ropes into their boats and remove the mussel clusters, reattaching any that are too small so that they can continue to grow.

The Tarbouriech Group: three generations and over six decades of aquaculture innovation

A resort and spa overlooking the Thau Lagoon in France, a boutique selling a line of bespoke cosmetics and numerous tasting places are among the services offered by Tarbouriech, a growing name in the Mediterranean hospitality sector. What consumers may not realize while enjoying a seaside lunch or relaxing at the spa, however, is that at the heart of this family-run enterprise is aquaculture.

In 1962, winegrower Pierre Tarbouriech left the vineyard for the lagoon to take on a new venture: oyster farming. Equipped with dedication and a passion for his craft, he grew the business alongside his family, introducing his son, Florent Tarbouriech, to the trade once he turned sixteen. Over the next 24 years, the Tarbouriech family, with Pierre at the helm, expanded the farm to include a 50 m² production house and three production tables on the lagoon.

Tragically, Pierre passed away in 1986 and the responsibility of continuing the family's legacy fell to 20-year-old Florent. Three years later, brimming with the same passion as his father and supported by his family, Florent established Médithau, the branch of the Tarbouriech Group encompassing the enterprise's shellfish farming activities, with the aim of sustainably providing healthy, high-quality seafood for everyone.

Within ten years of Florent's leadership, Tarbouriech became one of the leading producers of mussels and oysters in Europe, certified according to the highest international standards. This success was largely due to the enterprise's continued dedication to innovation and diversification.

Traditionally, in the Thau Lagoon, strings of mussels are suspended under tables and submerged into the water. Though the lagoon's waters already lends themselves to generously sized and great-tasting mussels, Tarbouriech strove to produce an even higher-quality product. In 1989, Tarbouriech partnered with researchers and other entities in the region to conduct tests on raising mussels in the Mediterranean using subsurface lines. These tests were deemed a success and for over ten years, Tarbouriech produced mussels on longlines at sea off the coast of Sète, France. Production boomed, with consumers impressed by the exceptional texture and flavour of the Tarbouriech mussels.

In the early 2000s, strong winter storms devastated the mussel farming structures. This damage, coupled with high levels of seabream predation, collapsed the volume of mussel production. Not to be deterred, the Tarbouriech family sought innovative solutions to re-establish production and preserve the traditional and historical practice of mussel farming in the region. By 2017, Tarbouriech installed storm-resistant subsurface lines, which had already proven effective in Italian waters, and adopted a work and maintenance philosophy to manage seabream populations. Thanks to this system,

annual mussel production has grown to 800 tonnes. Tarbouriech has also built an ultra-modern treatment centre dedicated to the processing of mussels, consisting of 3 500 m² of treatment rooms and laboratories and 1 750 m² of pools. Here, in addition to the 800 tonnes of mussels produced in house, the Group works to process an additional 5 000 tonnes of mussels from external producers annually.

In a further demonstration of their dedication to innovation, the Group is updating traditional Thau Lagoon farming techniques for another kind of shellfish: oyster. Florent Tarbouriech developed and patented a system that reproduces the tidal currents that are otherwise lacking in the Mediterranean Sea. This automated and remotely controlled system is powered by solar and wind energy and allows the Group to produce 350 tonnes of exceptional-quality oysters. This system has proven so effective that Tarbouriech has begun to venture beyond French waters and implement it in lagoons in Italy, Spain and Japan.

Tarbouriech aims to diversify beyond aquaculture. Their drive to share their art of living has always been channeled through the production and sale of quality seafood products, but it gained new momentum in 2011 when Florent's children Florie and Romain joined the Tarbouriech Group's

"Our mission is to sustainably supply the ever-increasing population with quality protein and to share our art of living."

adventure and created Saint-Barth, a restaurant at the water's edge serving freshly harvested Tarbouriech mussels and oysters – a second tasting place, Saint-Pierre, would be opened ten years later in Loupian with a picturesque view of Mont Saint-Clair. Since then, Tarbouriech has continued to expand into the field of ecotourism, opening the Domain Tarbouriech in 2018, a resort located in an ancient folly nestled between sea and lagoon where guests can relax, swim, enjoy the spa and dine on fresh shellfish at the on-site restaurant. The spa is stocked with products from Tarbouriech's own line of wellness products that are enriched with the power of mother-of-pearl from the Group's farmed oysters and marine active ingredients. In 2022, Tarbouriech continued its diversification efforts with the acquisition of a fish shop at the Jacques Coeur Hall in Montpellier and the opening of a sea bar offering Tarbouriech mussels and oysters and fresh fish direct from the Sète auction. The Group is currently pursuing the

development of a range of gluten-free products made with oyster meat for consumers to enjoy at home.

Tarbouriech has identified the need to address three key challenges to its operations: global warming, which is leading to anoxia of the Thau Lagoon environment and the death of marine organisms; the predation of oysters; and an increase in tourist activities, which can discharge wastewater into the environment, leading to closures and sales bans. In an effort to mitigate these threats, the Group is currently working alongside both private and public research centres.

With an impressive array of products to its name, 100 production structures located around the world and 1 150 tonnes of annual production, where does the Tarbouriech Group go next? Adhering steadfastly to their mission since the beginning, they aim to continue to sustainably supply consumers with high-quality protein and to share their art of living through continued development and new restaurants, hotels, spas and tasting places.

©GFCM/Daniel Gillet

"In Greece, fish is a very big part of the everyday diet and mussels are very popular. A traditional Greek way to prepare mussels is in a form similar to an Italian *risotto*, called *midopilafo*. The dish is made of toasted rice, slowly cooked with onions, white wine and stock before the mussels are added."

Vasileios Konstantinidis, Culinary Student at Institut Paul Bocuse

Traditions and recipes from the region

The consumption of mussels has a long history in the Mediterranean and Black Sea region. In the first century CE, Roman gastronomist and author of *De Re Coquinaria*, Marcus Gavius Apicius, described a recipe for mussels marinated in sweet wine and a fermented fish sauce called *garum*. Today, mussels are frequently served steamed, deep-fried or pan-fried in seaside restaurants or by street food vendors.

Italy / *Zuppa di cozze*

Prepared in the region around Naples, this recipe, invented in the eighteenth century, is traditionally served for Easter celebrations on Holy Thursday. Mussels are steamed with white wine and laurel leaves until the shells open and the remaining cooking liquid is filtered to remove any pieces of shell and set aside. Chopped celery, carrots and onions are sauteed in a pan with olive oil and the reserved cooking liquid is added along with fish broth and diced tomatoes. The soup is brought to a simmer and seasoned with black pepper and lemon juice. It is served in a bowl together with the steamed mussels.

Spain / *Tigres o Mejillones rellenos*

This recipe belongs to the famous class of tapas dishes served in Spain. The mussels are sorted and cleaned under fresh water, before being steamed to open their shells. Finely diced onions, tomatoes and green and red bell peppers are sauteed in a pan with a smoked paprika called *pimenton de la vera*. The mussels are garnished in their shells with the vegetable preparation and bechamel sauce before being breaded and oven-baked until golden-brown.

Türkiye / *Midye tava*

Served as street food along Turkish coastlines, steamed mussels are removed from their shells and put on skewers. The mussels are then battered and deep-fried in a huge metal cauldron. The mussel skewers are served with a tangy tarator sauce made from bread soaked in water that is blended with garlic, sesame paste and lemon juice until a smooth paste is formed. In some versions, poached mussels are added to the sauce to intensify the flavour.

Chef's tips

How to prepare mussels?

Selecting fresh mussels is crucial to prevent food poisoning. Even if they have been caught the same day, it is important to check each mussel individually and remove any that are open and do not react when manipulated as well as any with cracked shells. Clean the mussels by eliminating the small threads, called beards, and rinse them with fresh water before cooking.

How to cook mussels?

The ideal way to prepare mussels is by steaming them with a little liquid at a high temperature. This cooking method opens the mussels and prevents them from becoming chewy. Raw mussels are also a delicacy found in higher-end restaurants, but the freshness must be absolutely guaranteed to prevent food poisoning.

How to season mussels?

Raw or steamed mussels are enhanced by the acidity found in white wine, citrus or vinegar. Commonly paired with shallots and fresh herbs, they can also be prepared with richer aromatic ingredients, such as mushrooms or garlic confit. To intensify the mussel flavour in recipes, remove the shells and pan-fry them with olive oil in a very hot skillet.

Here are the best ingredients to pair with mussels for unique Mediterranean and Black Sea-inspired dishes

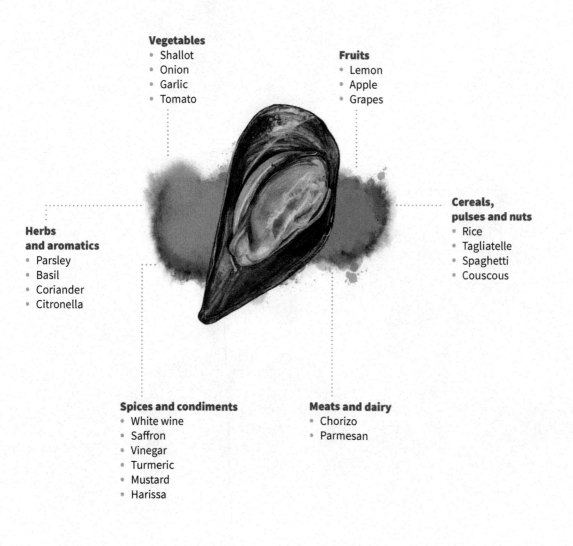

Vegetables
- Shallot
- Onion
- Garlic
- Tomato

Fruits
- Lemon
- Apple
- Grapes

Herbs and aromatics
- Parsley
- Basil
- Coriander
- Citronella

Cereals, pulses and nuts
- Rice
- Tagliatelle
- Spaghetti
- Couscous

Spices and condiments
- White wine
- Saffron
- Vinegar
- Turmeric
- Mustard
- Harissa

Meats and dairy
- Chorizo
- Parmesan

Mussel skewers with rosemary and pomegranate sauce

Type
starter
Yield
4 servings
Preparation time
20 minutes
Cooking time
10 minutes

This recipe is inspired by the mussel skewers found in Türkiye. The mussels are first steamed with white wine and garlic, then spiked on a rosemary stem and glazed with a tangy pomegranate–sumac sauce.

Ingredients

Utensils needed
- Cutting board
- Oyster knife
- Large pan
- Blender
- Fine sieve
- Saucepan
- Scissors
- Brush
- Baking tray

Steamed mussels
- 500 g mussels
- 1 white onion
- 2 garlic cloves
- 2 rosemary stems
- 1 glass of white wine (optional)

Pomegranate glaze
- 2 pomegranates
- Juice of 1 orange
- 1 pinch sumac
- 1 pinch cumin powder
- Salt and pepper

Mussel skewers
- 12 rosemary stems
- 4 pinches zaatar

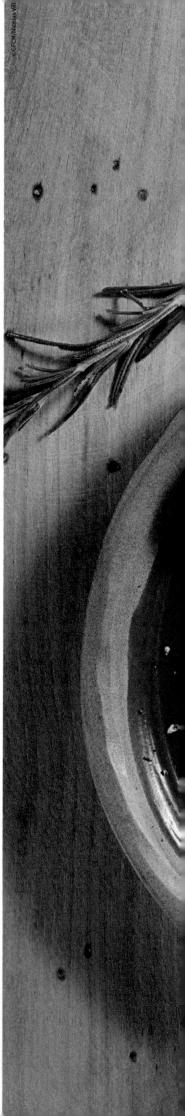

Preparation

Steamed mussels
Rinse the mussels under cold water and discard the broken ones. Peel and chop the onion and the garlic. In a large pan, sauté the onion and garlic in olive oil. Add the mussels, rosemary stems and wine, if using. Cover and simmer for 5 minutes until the mussels open. Remove the mussel flesh from the shells and refrigerate.

Pomegranate glaze
Cut the pomegranates in half and remove the seeds. Blend the orange juice and pomegranate seeds and pass the mixture through a fine sieve. In a saucepan, simmer the pomegranate and orange juice until it becomes a syrupy glaze. Season with the sumac, cumin, salt and pepper.

Mussel skewers
Cut each rosemary stem to 10 cm in length and remove the leaves from two-thirds of the stem. Use the rosemary stems as skewers with six to eight mussels per stem and glaze with the pomegranate–orange preparation using a brush. Bake the mussel skewers in an oven preheated to 180 °C for 5 minutes and finish with a sprinkle of zaatar.

Nutrition facts

	Per 100 g	Per recipe
Energy	283 kJ/68 kcal	3 397 kJ/812 kcal
Protein	5.2 g	62.1 g
Carbohydrate	6.9 g	82.7 g
Fibre	1.2 g	14.8 g
Sugar	4.5 g	54.1 g
Fat	1.2 g	13.8 g
Saturated fat	0.3 g	3.2 g
Sodium	173 mg	2 079 mg

Gilthead seabream

Sparus aurata

Gilthead seabream in the Mediterranean and Black Sea region

Scientific name:
Sparus aurata

Family:
Sparidae

Approximate average production volume (2016–2020):
227 900 tonnes

The top three producers are **Türkiye**, **Greece** and **Egypt**.

84% of production occurs in **marine water** and **16%** in **brackish water**.

Popular among chefs in the region for its **versatility, mild flavour** and **health benefits**.

Source: FAO. 2023. Global aquaculture production quantity (1950–2020). In: *Fisheries and Aquaculture Division*. Rome. Cited 26 July 2022. fao.org/fishery/statistics-query/en/aquaculture/aquaculture_quantity

Based on data from the GFCM Information System for the Promotion of Aquaculture in the Mediterranean (SIPAM)

Sparus aurata

Gilthead seabream

T he gilthead seabream (*Sparus aurata*) is the only member of the Sparidae family that is commercially produced on a large scale in the Mediterranean and Black Sea region. The name "aurata" derives from the peculiar shiny golden frontal band between the eyes, very well recognizable in adult specimens. It is commonly found throughout the Mediterranean basin and its reach can occasionally extend to the Black Sea (Pavlidis and Mylonas, eds., 2011). Thanks to its good market price, high survival rate and feeding habits, gilthead seabream is considered well suited to marine cage aquaculture and onshore, recirculating aquaculture systems and is a popular choice among farmers in the region.

Culinary and nutritional value

Seabream is a very popular fish known among chefs around the Mediterranean and Black Sea region for its versatility. Served whole, either grilled on an open fire, pan-fried or steamed, seabream makes for an ideal meal, as its size corresponds to a one-person serving. Its delicate flesh with mild flavours is perfect both raw and cooked, including as tartare, pan-fried fillets or grilled whole on the barbecue. Its low-fat and high-protein content places seabream among the fish that can be healthily consumed on a weekly basis.

Farming in the region

Gilthead seabream is produced throughout the Mediterranean region, reaching a volume of approximately 265 900 tonnes in 2020, with Türkiye, Greece and Egypt as the largest producers (FAO, 2023). Traditionally, the culture of gilthead seabream took place in extensive systems that capitalized on the natural migration of juveniles to coastal lagoons (Seginer, 2016). But by 1981, faced with difficulties in finding enough wild juveniles to maintain an increasing demand, Italian producers had begun to successfully breed the species

artificially and by the end of the decade, attempts were being made in Spain, Italy and Greece to produce juveniles on a large scale (FAO, 2022d). Today, gilthead seabream can be farmed extensively, semi-intensively or intensively, each characterized by increasing levels of human control over the farming system. Extensive production is based upon the natural migration of the fish, semi-intensive systems can involve either seeding coastal lagoons with prefattened juveniles or fertilizing the area, and intensive systems – the method with the most human control – involves stocking land-based installations or sea cages with juveniles obtained from

hatcheries. In the Mediterranean, intensive ongrowing in sea cages is common; despite the inability to control temperature and the resulting extension of the growing period, it is the simplest and most cost-effective method of fattening (FAO, 2022d). The market size of the whole fresh fish in supermarkets is around 350–400 g, corresponding to one portion size.

Al-Bahar: adversity breeds resilience, and fish

Aquaculture is emerging as a crucial industry supporting food security in Gaza. With the help of funds from the United States Agency for International Development (USAID) and the Arab Reconstruction Authority, fish farms like Al-Bahar in Gaza are increasing production, improving quality and shaking off their reliance on imports.

Abu Yazan cofounded Al-Bahar in 2015 with his colleagues from the Mahmoud Al-Haj Sons company. Abu Yazan had made several recent trips to the United Arab Emirates and the country's progress on aquaculture inspired him to try to establish an economically viable farm in Gaza. Gilthead seabream, one of the most popular seafood species in the region, was chosen as the basis for the project, complemented by a smaller volume of seabass production.

When constructing the business, Abu Yazan envisioned an interactive, tank-to-table experience for customers, with a seaside restaurant built within a stone's throw of the same fish tanks that would be providing patrons their lunches. Today it has become the farm's tradition to invite families to take a stroll through the farm before their meal and to peer into the tanks to select the fish they would like to eat.

High demand for Al-Bahar's products from restaurant-goers, home cooks and a Gaza market dealer led the farm's ownership to launch an expansion project in 2017. Eighteen new tanks, made of metal and covered by sheets of polyvinyl chloride, were built about 100 m south of the original 15, and in 2018, construction of the farm's own hatchery was completed.

At the same time, existing constraints in the Gaza region have created many challenges for Al-Bahar. The territory continues to experience frequent electricity outages, which present a serious problem for aquaculture farms that need to provide their fish with consistently oxygenated water around the clock. Three generators had been installed to provide backup in such cases of emergency, but their costs of operation were high and technical issues led to several mass mortality events at the farm. These financial strains added to the burden of purchasing costly fish feed and – before construction of the hatchery – eggs and fry.

In 2018, a USAID initiative to supply farmers in Gaza with renewable energy sources, independent of the existing grid, gave a boost to Al-Bahar's chances for sustained aquaculture production. A 100 kW system of solar panels was installed at the farm through a grant from USAID. The Arab Reconstruction Authority supplied an additional 100 kW solar energy system a year later and by 2022, another 550 kW had been added, bringing the farm's annual electricity bill down from USD 50 000 to USD 11 000 in just four years.

The extra money freed up by these initiatives has allowed Al-Bahar to diversify and innovate in its production methods and achieve a higher-quality product. Instead of releasing 30 000 imported fry into a tank and rearing them together throughout the production process, as the farm's team had done in the early days, they now place 100 000 hatchery-bred fry into a single tank, from which maturing juveniles are separated into new tanks based on growth rates. This grouping of fish with similar life history trajectories allows for careful calibration of the heat supplied to each tank and optimal use of water volume to maximize output. In 2018, Al-Bahar began to sell its surplus fish to the West Bank; today, 75 percent of the seabream produced is sent there.

Thanks to the work of Abu Yazan and the entire Al-Bahar team, on this sandy stretch along the beachfront boulevard Al-Rasheed Street, with its farm-raised fish, children's playground and two-story seafood restaurant, a peacefulness settles in and resilience, self-reliance and ingenuity are on full display.

"Seabream is one of the most important fish consumed in Türkiye. This fish can be served inside a sandwich, as in the case of *balık-ekmek*, which is filled with grilled seabream, parsley and chopped onion."

Dilara Cimen,
Culinary Student at Institut Paul Bocuse

©GFCM/Daniel Gillet

"The aim from the beginning was to sustainably supply the local market with large quantities of fish."

Traditions and recipes from the region

Seabream can be prepared in various culinary applications, from grilled on a stick over an open fire to poached in a spiced broth. An ancient Roman recipe mentions a technique of baking seabream wrapped in fig leaves in a terracotta pot. Today, seabream is often served raw in restaurants as a tartare or as a carpaccio seasoned with citrus and fresh herbs.

Cyprus / **Baked seabream with ouzo**

Also called zoukki in Cyprus, the aniseed alcohol better known as ouzo is used in the preparation of a traditional seabream recipe. The fish is scaled, gutted and stuffed with slices of lemon, wild fennel, dill, rosemary, oregano and garlic. It is then drizzled with olive oil and ouzo. Depending on the recipe, the seabream is then placed on an oven dish with slices of potatoes, fennel or tomatoes and baked until the inner flesh is cooked and the skin is crispy.

Italy / *Orata alla pugliese*

The traditional way of preparing seabream in the southern Italian region of Puglia is to bake it between layers of potato slices seasoned with parsley, garlic, lemon and white wine. In an oven dish, a first layer of finely sliced potatoes is arranged and seasoned with chopped parsley, garlic and grated pecorino. The seabream is placed on top and covered with a second layer of seasoned potatoes. Olive oil is drizzled all over before the dish is baked in a preheated oven until the potatoes are cooked through. Covering the fish with potatoes prevents the flesh from overcooking and becoming dry.

Türkiye / *Izgara Çipura*

In this traditional Turkish recipe, seabream is gutted and scaled, then the skin is cut with shallow incisions so that it does not tear during cooking. Garlic and black pepper are crushed with salt and olive oil in a mortar and spread over the seabream with a basting brush. The seabream is then grilled on an open fire until the skin turns golden-brown. For this process, grill tongs can be used to prevent the skin from sticking to the grill. The grilled seabream is served with a salad or grilled vegetables.

Chef's tips

How to prepare seabream?

When preparing raw seabream tartare, it is important to scale the fish and remove the fins to prevent it from slipping when extracting the fillets. Use scissors to remove the fins, trying to follow the fillet line as closely as possible. The raw fish must be consumed within a few hours of preparation and stored in the fridge.

How to cook seabream?

Seabream is the perfect size for an individual portion, making it ideal for the grill, pan-frying or baking and serving whole with a sauce. The white flesh of the seabream is fragile, so high-heat treatment for a short amount of time should be favoured over poaching or preparing it in a stew.

How to season seabream?

Seabream has a mild flesh that pairs well with aromatic herbs and citrus. When used raw, the fish matches perfectly with fresh and acidic ingredients including pomegranate, grapefruit and fresh tomatoes. When grilled or pan-fried, intense aromatic ingredients such as mushrooms and saffron can be used to flavour the flesh.

Here are the best ingredients to pair with seabream for unique Mediterranean and Black Sea-inspired dishes

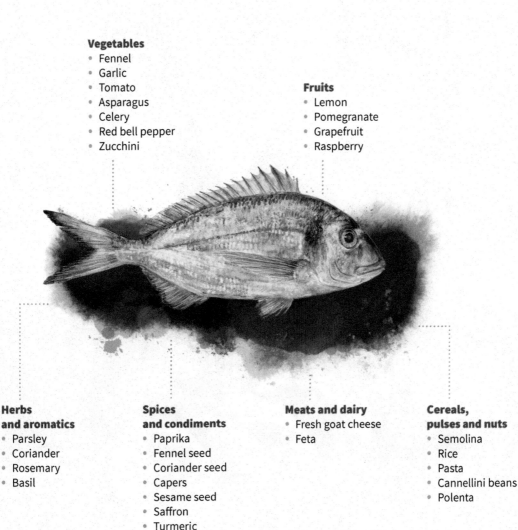

Vegetables
- Fennel
- Garlic
- Tomato
- Asparagus
- Celery
- Red bell pepper
- Zucchini

Fruits
- Lemon
- Pomegranate
- Grapefruit
- Raspberry

Herbs and aromatics
- Parsley
- Coriander
- Rosemary
- Basil

Spices and condiments
- Paprika
- Fennel seed
- Coriander seed
- Capers
- Sesame seed
- Saffron
- Turmeric

Meats and dairy
- Fresh goat cheese
- Feta

Cereals, pulses and nuts
- Semolina
- Rice
- Pasta
- Cannellini beans
- Polenta

Barbecued seabream stuffed with green fennel, orange and capers

Type
main course
Yield
4 servings
Preparation time
20 minutes
Cooking time
15 minutes

Inspired by the traditional Mediterranean method of grilling fishes on an open fire or on charcoal, this recipe showcases the process of stuffing and grilling seabream on a barbecue. The fish is paired with fennel, bringing aniseed notes, and orange, giving a delicate citrus perfume to the seabream.

Ingredients

Utensils needed
- Cutting board
- Non-stick pan
- Mixing bowl
- Baking tray
- Greaseproof paper
- Scissors
- Fish bone tweezers
- Saucepan
- Blender
- Plates

Orange and garlic sauce
- 3 tbsp extra virgin olive oil
- Zest and juice of
  2 organic oranges
- 1 glass of white wine (optional)
- 1 garlic clove
- 1 tbsp honey
- Salt and pepper

Almond and caper topping
- ½ jar of capers
- 1 handful of slivered almonds

Barbecued seabream
- 4 seabreams
- 1 tbsp extra virgin olive oil
- 1 fennel bulb
- 2 organic oranges
- Fine salt

Preparation

Orange and garlic sauce
Peel and chop the garlic. In a saucepan, cook the orange juice, zest, garlic, olive oil, white wine, if using, and honey over low heat for 5 minutes. Blend until smooth and season with salt and pepper.

Almond and caper topping
Roast the almonds in an oven preheated to 180 °C for 5 minutes. In a small pan, fry the capers over high heat until crispy.

Barbecued seabream
Remove the scales, fins and gills of the seabream and rinse under cold water. Cut shallow incisions in the skin to prevent it from cracking while cooking. Finely slice the oranges and fennel, stuff them in the seabream and season with fine salt. Grill on a barbecue for 2 minutes on each side, until the skin is golden brown and crispy.

To serve
Place each seabream in the centre of a plate. Top with slivered almonds and fried capers and serve the orange and garlic sauce on the side.

Nutrition facts

	Per 100 g	Per recipe
Energy	481 kJ/115 kcal	10 159 kJ/2 428 kcal
Protein	12.4 g	262 g
Carbohydrate	2.3 g	48.1 g
Fibre	0.6 g	11.5 g
Sugar	1.7 g	36.2 g
Fat	5.8 g	121.0 g
Saturated fat	1.0 g	21.1 g
Sodium	96.2 mg	2 029 mg

Flathead grey mullet

Mugil cephalus

Flathead grey mullet in the Mediterranean and Black Sea region

Scientific name:
Mugil cephalus

Family:
Mugilidae

Approximate average
production volume
(2016–2020):
320 900 tonnes

The top three producers are
Egypt, **Israel** and **Italy**.

78% of production
occurs in **freshwater** and
22% in **brackish water**.

An **affordable** and **healthy**
fish, perfect for both everyday
cuisine and festive occasions.

Source: FAO. 2023. Global aquaculture production quantity (1950–2020).
In: *Fisheries and Aquaculture Division*. Rome. Cited 26 July 2022.
fao.org/fishery/statistics-query/en/aquaculture/aquaculture_quantity

Based on data from the GFCM Information System for
the Promotion of Aquaculture in the Mediterranean (SIPAM)

Mugil cephalus

Flathead grey mullet

Known for its hardiness, simple diet and rapid growth, flathead grey mullet (*Mugil cephalus*) has been both farmed and fished in the Mediterranean and Black Sea region for centuries. It often roams coastal waters and can enter estuaries, rivers and ports thanks to its adaptability to different salinity levels. Not only has the flesh of this fish been popular since the time of ancient Egypt, but its roe, known as *bottarga* or *poutarque*, has been considered a delicacy since the time of ancient Rome and is today enjoyed by a global market.

Culinary and nutritional value

Grey mullet's firm greyish flesh is rich in proteins, fatty acids and vitamin B₆ and can be pan-fried, deep-fried, grilled on an open fire or even smoked. Its affordable price and diverse uses make it an ideal fish for both everyday cuisine and festive occasions. Grey mullet roe is also used in the creation of a Mediterranean delicacy called *bottarga*, which is consumed on toast or grated onto sublime dishes in Italy, France, Egypt and Türkiye.

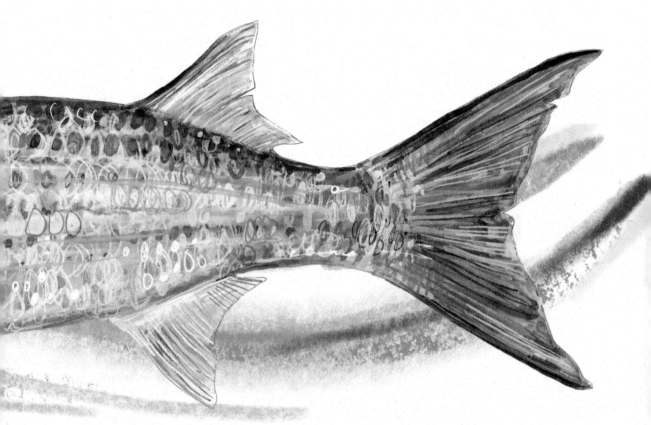

Farming in the region

Farming of flathead grey mullet dates back centuries in the Mediterranean region and to 1930 in the Black Sea region (Saleh, 2008; Saleh and Salem, 2005). Since then, production has expanded across numerous regions of the world.

While the production of many farmed species begins by producing juveniles, also known as fry, the production of most flathead grey mullet begins by collecting fry from the wild when the larvae migrate to inshore waters and estuaries. Once collected, the fry must gradually acclimatize to the salinity levels of the aquaculture system; if they do not, producers may see a 100 percent mortality rate over the following two weeks. Fry are then stocked in earthen nurseries, where they are kept for four to six months until they reach fingerling size, about 10 g (FAO, 2022e). The next phase of the aquaculture process involves the ongrowing of the fingerlings and lasts until they reach an acceptable commercial size. Flathead grey mullet are typically grown in a polyculture system, meaning they are stocked in ponds along with other fish, often including common carp (*Cyprinus carpio*), silver carp (*Hypophthalmichthys molitrix*) and tilapia (*Oreochromis* spp.)

(Saleh, 2008). This process takes approximately seven to eight months, after which time the fish reach 0.75–1 kg; depending on market demand, the fish may be reared for an additional ongrowing season until they reach 1.5–1.75 kg (FAO, 2022e). In ponds or nets, producers can harvest daily, based on market demand (FAO, 2022e). Flathead grey mullet is also an excellent candidate for the ancient aquaculture practice of valliculture. This farming technique involves taking advantage of the migration patterns of fish by preventing them from returning to the sea once they reach a lagoon or brackish water body.

Mohamed Souei:
no compromises between sustainability and quality

Along the Tunisian coast, 10 km from the Libyan border, lies a narrow sand bar. Off one side stretches the Mediterranean Sea, off the other the rich, protected waters of the El Bibane Lagoon. This 33 km long lagoon is the second largest in Tunisia and among the largest in the Mediterranean basin. Its 25 000 ha are home to a unique ecosystem and a remarkable variety of flora and fauna that have earned it the title of a wetland of international importance and a place on the Ramsar list.

El Bibane has also proven to be of great social and economic importance for the region. For centuries, fishers and farmers have coexisted, providing themselves, their families and their communities with healthy and high-quality aquatic foods. Among these farmers is Mohamad Souei.

Souei is a representative of the latest generation to steward the El Bibane Lagoon and its resources. He approaches his work with a passion for aquaculture and a deep respect for the environment. It was this passion that drove him to leave his previous profession in 2006 to become an entrepreneur in the aquaculture sector. His dream was to establish a business dedicated to fusing sustainability and quality to create a highly appreciated farmed aquatic product. Souei began work towards realizing this dream by farming seabass and seabream in cages at an intensive off-shore farm in Ben Guerdene, Tunisia, located west of the El Bibane Lagoon. Since then, he has shifted his operations to El Bibane, where he works with a team of 38 and collaborates with local stakeholder to preserve the lagoon.

Souei prioritizes minimizing his environmental impact, opting for extensive farming practices that rely on natural feed and ancient techniques. He focuses on the culture of flathead grey mullet, a member of the Mugil genus and the most captured species in the lagoon due to its strong reproduction and growth supported by the water current and salinity.

In the El Bibane Lagoon, the culture process begins in the autumn and winter with the natural recruitment of fingerlings. Souei captures them in the lagoon, one of the few times he opts to intervene in the natural life cycle of the fish. To catch the flathead grey mullets during their migration, a special trapping system called a *bordigue* is employed. This masterpiece in design dates to the fourteenth century and is one of the oldest aquaculture techniques used in Tunisia and in the wider Mediterranean region. Souei's *bordigue* is made of a fixed barrier measuring more than 3.6 km, along with thirty-nine capture chambers – three sections of the barrier with five capture chambers and eight sections with three capture chambers – each referred to as a "catch room". Once the fingerlings have been recruited in the lagoon, Souei allows them to grow naturally and is able to

©GFCM/Sahbi Dorai

harvest the fish as needed. His reliance on natural methods and the ancient *bordigue* method helps Souei preserve the freshness and high quality of the fish and allows him to maintain active production approximately 300 days out of the year.

In association with administrations, researchers and fishers, Souei works hard to maintain a balanced and healthy ecosystem in the lagoon. He rigorously adheres to regulations, including by avoiding prohibited zones and operating row boats to minimize the environmental impacts of his activities.

In 2022, he launched a new initiative in the lagoon, catch and release fishing, which allows passionate sport fishers to fish in the area while maintaining its biodiversity. This initiative allowed him to support the local economy through the creation of new jobs, particularly for women who prepare meals for the tourist fishers.

Souei explains that regardless of future challenges and changing circumstances, such as those faced during the height of the COVID-19 crisis when production in the region decreased and farmers' incomes fell, he will continue to dedicate himself to producing a high-quality product appreciated by consumers while maintaining respect for the environment.

"The idea is to maintain a business with respect for the environment and for the quality of the final product."

"Fish, and in particular flathead grey mullet, is a big part of the diet in Israel. Traditionally, the fish is included in *chraime*, a dish of flathead grey mullet simmered in a sauce of olive oil, harissa, caraway and cumin and served alongside bread."

Dor Gali, Culinary Student at Institut Paul Bocuse

Traditions
and recipes from the region

S ought after since antiquity for its roe, flathead grey mullet is a long-standing pillar of the Mediterranean gastronomic tradition. Historically, the fillets of the fish were commonly cold smoked to help with preservation and to develop their sensorial characteristics. Contrary to its more recent reputation as a low-quality "harbour fish", flathead grey mullet produced in aquaculture farms is mild in flavour and ideal when prepared pan-fried or cured.

Egypt / *Fesikh*

Eaten during the Egyptian spring festival of *Sham Ennessim*, this culinary specialty is prepared using grey mullet that are salted, fermented and dried in the sun. If not performed correctly, the fermentation process can cause potentially lethal poisoning. Therefore, *fesikh* should be bought in specialized stores and kept in firmly closed glass jars. The mullet is accompanied by finely chopped onion, lemon wedges and an Egyptian flatbread called *ayesh baladi*.

Italy / *Bottarga*

This Mediterranean delicacy from Sardinia is made from grey mullet roe that is cured in salt, pressed, then dried and covered with a fine layer of wax to preserve its sensorial qualities over time. *Bottarga* is served in fine slices on warm toast or grated on top of pasta dishes. The deep and complex aromatic notes transform a simple dish into a gastronomic experience.

Spain / *Mujol a la sal*

This recipe from the Menor Sea is made of scaled and gutted grey mullet. The fish is covered with coarse salt and baked in a preheated oven. The salt crust keeps the moisture of the grey mullet inside and at the same time seasons the flesh. This dish is served with a traditional Spanish mayonnaise sauce seasoned with garlic called *salsa de ajos*. Some versions of this sauce can also be made with chopped parsley or coriander.

Chef's tips

How to prepare grey mullet?

To prevent grey mullet flesh from becoming dry during cooking, fillets can be soaked in a brine prepared with 1 litre of ice water and 50 g of coarse salt. Mix the salt into the water until dissolved, before adding the fillets and leaving them to soak in the fridge for 20 minutes. Then, rinse the fillets under cold water and wipe off the excess moisture with a clean towel.

How to cook grey mullet?

Grey mullet's smaller size facilitates its use as a whole fish, either pan-fried, grilled or even deep-fried. The gills and the guts should be removed prior to cooking as they can add bitter notes to the flesh. A great way to prepare grey mullet is to butterfly it by removing the head, cutting open the underside and removing the bones. The fillets can then be breaded and pan-fried on both sides.

How to season grey mullet?

Grey mullet is an affordable fish that can be prepared in a myriad of ways, including simply grilled with lemon juice or grilled with a spicy sauce. Compared to more expensive fishes, like seabass or turbot, grey mullet offers an ideal basis to experiment with uncommon ingredient pairings, such as an emulsified sauce of tarragon, tangerine and olive oil.

Here are the best ingredients to pair with flathead grey mullet for unique Mediterranean and Black Sea-inspired dishes

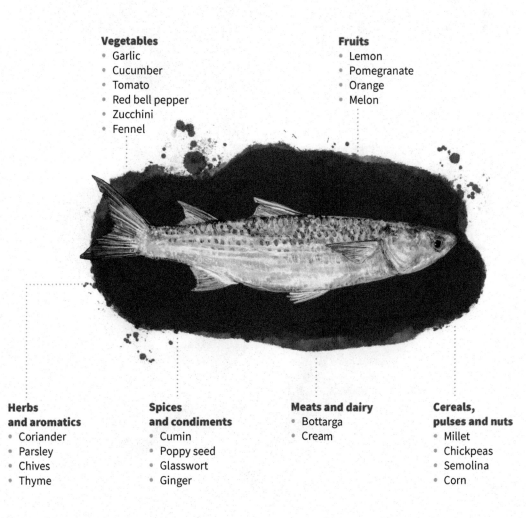

Vegetables
- Garlic
- Cucumber
- Tomato
- Red bell pepper
- Zucchini
- Fennel

Fruits
- Lemon
- Pomegranate
- Orange
- Melon

Herbs and aromatics
- Coriander
- Parsley
- Chives
- Thyme

Spices and condiments
- Cumin
- Poppy seed
- Glasswort
- Ginger

Meats and dairy
- Bottarga
- Cream

Cereals, pulses and nuts
- Millet
- Chickpeas
- Semolina
- Corn

Butterflied mullet with polenta breading and tarragon aioli

Type
main course
Yield
4 servings
Preparation time
20 minutes
Cooking time
30 minutes

This dish showcases the technique of butterflying, breading and pan-frying round fishes such as flathead grey mullet, replacing breadcrumbs with polenta in a Slovenian-inspired twist. The flathead grey mullet is paired with candied cherry tomatoes and an herbaceous tarragon mayonnaise.

Ingredients

Tarragon aioli
- 2 garlic cloves
- 1 egg yolk
- ½ bunch of tarragon
- ¼ bunch of parsley
- Juice of ½ lemon
- 4 tbsp extra virgin olive oil
- Salt and pepper

Candied cherry tomatoes
- 1 handful of yellow cherry tomatoes
- 1 tbsp extra virgin olive oil
- 1 pinch fine salt
- 2 pinches white sugar
- 1 sprig of fresh thyme
- 2 garlic cloves

Breaded mullet
- 1 flathead grey mullet
- Salt and pepper
- ½ tbsp oregano
- 150 g flour, type 55
- 2 eggs
- 150 g polenta
- 5 tbsp extra virgin olive oil

Utensils needed
- Garlic press
- Blender
- Cutting board
- Baking tray
- Greaseproof paper
- Scissors
- Fishbone tweezers
- Shallow dishes
- Non-stick pan
- Mixing bowl

Preparation

Tarragon aioli
Peel and press the garlic and blend with the egg yolk, tarragon, parsley and lemon juice until smooth. Gradually pour in the olive oil while blending until emulsified. Season with salt and pepper.

Candied cherry tomatoes
Wash and cut the cherry tomatoes in half. Place them on a baking tray lined with greaseproof paper, sprinkle with olive oil, salt and sugar. Top with sprigs of thyme and garlic cloves and bake in an oven preheated to 160 °C for 20 minutes.

Breaded mullet
Rinse the mullet under water and remove the scales, gills and fins. Butterfly the fish by removing the spine from the belly while keeping the fillets attached together at the back. Remove the bones from the fillets and season with salt, pepper and oregano. In three separate shallow dishes, add flour, beaten eggs and polenta. Coat the fish in flour, followed by the egg and finally the polenta. Over medium heat, pan-fry the breaded fish in olive oil for 6 minutes on both sides, until golden-brown.

To serve
Place the fish in the centre of a dish and top with the candied cherry tomatoes and tarragon aioli.

Nutrition facts

	Per 100 g	Per recipe
Energy	283 kJ/68 kcal	3 397 kJ/812 kcal
Protein	5.2 g	62.1 g
Carbohydrate	6.9 g	82.7 g
Fibre	1.2 g	14.8 g
Sugar	4.5 g	54.1 g
Fat	1.2 g	13.8 g
Saturated fat	0.3 g	3.2 g
Sodium	173 mg	2 079 mg

Stony sea urchin

Paracentrotus lividus

Stony sea urchin in the Mediterranean
and Black Sea region

Scientific name:
Paracentrotus lividus

Family:
Parechinidae

Approximate average
production volume
(2016–2020):
Commercial production has
yet to begin in the region

Produced in **tanks**, **baskets**
and **cages**.

Consumers' appreciation
for this species has inspired
producers to progress with
exciting innovations
in the region.

Source: FAO. 2023. Global aquaculture production quantity (1950–2020).
In: *Fisheries and Aquaculture Division.* Rome. Cited 26 July 2022.
fao.org/fishery/statistics-query/en/aquaculture/aquaculture_quantity
Based on data from the GFCM Information System for
the Promotion of Aquaculture in the Mediterranean (SIPAM).

Paracentrotus lividus

Stony sea urchin

T hough it has been harvested for thousands of years in the Mediterranean basin, commercial aquaculture production of the stony sea urchin (*Paracentrotus lividus*) has yet to develop in the region. This spiny, round echinoderm has been defined as one of the most important herbivores in the region, both as an indicator of a healthy environment and a means to control algal overgrowth. Consumers' great appreciation for this product has inspired producers to progress with exciting innovations that could open up aquaculture of the species to the valuable Mediterranean market in the near future.

Culinary and nutritional value

Sea urchin is an exceptional product with an incomparable taste emblematic of the Mediterranean region. It is consumed on special occasions and served in gourmet restaurants. Within sea urchin's complex, iodine flavour, it is possible to differentiate sweet notes including those of dried fruit and hazelnut. It is rich in protein and minerals and very low in fat. The gonads, also known as roe, can be consumed raw on toast or with eggs or used in the preparation of risotto and fresh pasta.

Farming in the region

While the stony sea urchin is widespread throughout the Mediterranean Sea (Boudouresque and Verlaque, 2007), commercial culture of the species has yet to be established in the region. Meanwhile, other countries have been farming other sea urchin species for decades. Globally, the most significant volumes of production are tied to Japan, where the first farming of sea urchins began in 1968, and China, where sea urchins have been cultured since the 2000s (Liu and Chang, 2015). The production process of the species differs depending on the country and important research is underway to optimize production both for commercial purposes and restocking activities. Generally, urchin farming revolves around spawning and rearing in land-based hatcheries, as well as gonad enhancement. In numerous countries, adult broodstock are induced to spawn and the resulting larvae are fed a diet of algal species until they reach an appropriate size. The long ongrowing process, lasting from 2.5 to 4 years, takes place in tanks, cages or baskets (McBride, 2005; Brundu et al., 2020). The resulting sea urchins can be supplied to the commercial market or to aid in wild stock enhancement of the species. As research continues, new culture methods, in particular those tailored to the specificities of the Mediterranean and the Black Sea, will likely emerge.

Urchinomics: turning barrens into boons

You've heard of economics and genomics, Reaganomics and Freakonomics – now get ready for Urchinomics! In 2021, Urchinomics launched the first and only viable land-based commercial urchin ranching facility in the world. Since the construction of this farm in Oita, Japan, Urchinomics has expanded its operations to Norway, Canada, the United States of America and the Mediterranean Sea. Across these regions, further commercial farms are under development and researchers have been engaged to study important aspects of the nascent industry, such as the effectiveness of feed and ranching technologies. The sites of Urchinomics' current and planned facilities map onto areas of highly depleted seaweed forests, known as urchin barrens, since urchin ranching and the restoration of ecologically vital seaweed forests go hand in hand.

In some areas, "overfishing, climate change and pollution have allowed sea urchins to explode in population, overgrazing entire kelp forests, seaweed beds and sea grass meadows and turning them into lifeless, desert-like barrens," writes Urchinomics founder and CEO Brian Takeda. Over the past decade, ocean warming effects have removed sea urchin predators and created a perfect storm of environmental conditions for urchin species, particularly the stony sea urchin, to proliferate prodigiously and reduce once vast forests of densely tangled kelp into carpets of hungry, spiky balls along the sea floor.

The urchins are hungry because once they have devoured all seaweed in sight, there is little left to feed their enormous populations. Inside the starving urchins, their fleshy roe – known by the name *uni* and prized as a delicacy in Japan and increasingly around the world – shrinks or disappears entirely, reducing the economic potential of harvesting the creatures and discouraging predators from eating them.

Urchinomics aims to break this vicious cycle by creating a financial incentive for fishers and divers to return to the same jungly waters they used to swim through in search of healthy, valuable urchins in order to now begin clearing the wastelands of empty balls. These malnourished urchins have little value when they come out of the water, but that's where the magic of Urchinomics begins. Once transported to the company's land-based recirculating aquaculture farm, the urchins are fed a formulated feed that brings them back to health and produces premium roe within just six to twelve weeks.

Near Barcelona, the stony sea urchin is exploding out of control. So, the GFCM has turned to Urchinomics, collaborating with the company on a field study in Spain. Research has shown that Urchinomics' technologies may

> "Our objective is to restore marine ecosystems by removing urchins from overpopulated areas and turning them into a premium seafood product."

"In Málaga, Spain, we eat seafood almost every day, typically as small dishes at lunch. Sea urchins can be prepared by stirring an urchin juice reduction, champagne and egg yolks into a bechamel sauce before adding the sea urchin roe and finally baking the entire dish in the oven."

Nuria Garrido, Culinary Student at Institut Paul Bocuse

well be able to convert the area's destructive stony sea urchins into a premium seafood product.

Hunting for urchins in barrens offers a fairly flexible part- or full-time occupation to fishers and divers. Empty urchins are much more abundant than urchins with roe and tend to be found closer to shore, and as the urchins will be nursed back to health on land, fishers need not worry about following the animals' biological calendars to harvest them at times of highest yield.

Parts of the Mediterranean Sea suffer from the opposite problem: too few urchins. For example, off the shores of the volcanic island of Procida, in Italy's Gulf of Naples, the natural controls keeping stony sea urchins from multiplying exponentially are still in place, but the population has been depleted to satisfy consumer demand. The stony sea urchin is an important herbivore in the region's ancient food chain and its health serves as an indicator in the assessment of environmental quality. In the region, there has been success in culturing stony sea urchins to rebuild these populations, starting with *in vitro* fertilization all the way through maturation and release for restocking into the wild.

The farming process is relatively straightforward. Since stony sea urchins are found low on the food chain, their nutritional needs are

not difficult to satisfy artificially. In Urchinomics' shallow raceways – narrow man-made channels that require little seawater and little energy to heat or cool the water to optimal conditions – the urchins are given a feed based on none other than their favorite meal: kelp. This kelp consists of the offcuts of sustainably produced kombu kelp for human consumption, also reducing food waste.

Urchin removal has been shown to bring seaweed communities back to life in as little as three to six months. And with the growing popularity of urchin roe as an ingredient in Mediterranean cuisine, a market for high-quality

urchins is sure to persist. The work of Urchinomics to transform this ecological problem into valued gourmet *uni*, in combination with efforts by farms in the region to buffer any harmful effects of this market by restocking depleted populations, is essential for the health of seaweed and marine ecosystems worldwide.

These actions – urchin removal and enhancement in overpopulated areas, and farming and restocking in areas with depleted populations – are also known as forms of restorative aquaculture, which can positively affect ecosystem services. Supporting restorative aquaculture practices such as these is integral to the GFCM 2030 Strategy.

Traditions
and recipes from the region

Sea urchin from the Mediterranean Sea has been consumed since antiquity. Its unique flavours make it highly appreciated in restaurants. Sea urchin roe freshly extracted from the shell is delicate and dissolves when cooked. It can be used raw added onto toast with drops of lemon juice or blended into sauces or soups to give them a distinct aromatic profile.

France / *Œufs brouillés aux oursins*

Traditional French cuisine provides a worldwide reference for egg recipes. One of the finest and most prized culinary applications is scrambled eggs served with urchin roe. The eggs are seasoned with urchin juice obtained by filtering the liquid from inside the shells once they are opened. The eggs are then delicately scrambled in a pan with butter. A spoonful of sour cream is added at the end to prevent the eggs from overcooking and becoming dry. The urchins' shells are cleaned and used as a bowl to serve the scrambled eggs, which are garnished with the urchin roe and fresh dill. This recipe can be served for breakfast or as a starter for an elegant dinner.

Greece / *Achinos taramosalata*

The traditional recipe for *taramosalata* (or *taramasalata*) is made with white fish, olive oil, lemon juice, grated onion and bread. Sea urchin roe may also be used to substitute the white fish. It is often served as part of a meze platter for special occasions. Take stale bread or sandwich bread and use only the crumbs – this will add texture to the recipe. Soak the bread in water and wring it out lightly, then blend all ingredients until smooth and creamy. The dish is ready to enjoy on toast or as a dip with vegetable sticks and is perfect as an appetizer.

Italy / *Spaghetti ai ricci di mare*

This pasta speciality from the Sicilian port city of Palermo is one of the gastronomic delicacies of the Mediterranean Sea. The sea urchins are opened to remove the roe. In the meantime, spaghetti is cooked in boiling salted water and finely chopped garlic and shallots are sauteed in a non-stick pan with olive oil. White wine is added to the pan and reduced to half its volume. Then the *al dente* spaghetti and sea urchin roe are tossed together. The pasta is served with chopped garlic.

Chef's tips

How to prepare sea urchin?

When preparing sea urchin, use thick leather gloves to protect your hand. Cut the spikes off with scissors, then incise the softer shell around the mouth while being careful not to poke the orange roe. Remove the liquid and the urchin's gut before gently detaching the roe from its shell with a spoon. Rinse the shell under cold water and wipe off the moisture with a clean towel.

How to cook sea urchin?

Sea urchin roe is a highly prized delicacy whose quality can easily degrade if prepared the wrong way. It is already a gastronomic experience to serve it raw on warm toast with drops of lemon juice. When incorporated in recipes, it needs to be added at the last minute of preparation. However, be careful seasoning with salt as the roe is already very salty.

How to season urchin?

Sea urchin roe is very complex aromatically and doesn't need any additional seasoning to express its full culinary potential. The rich iodine aromas elevate simple dishes such as scrambled eggs or fresh linguine to gastronomic heights. To intensify the flavour, the liquid inside the shell can be filtered and used in the preparation of a sauce or a bouillon.

Here are the best ingredients to pair with sea urchin for unique Mediterranean and Black Sea-inspired dishes

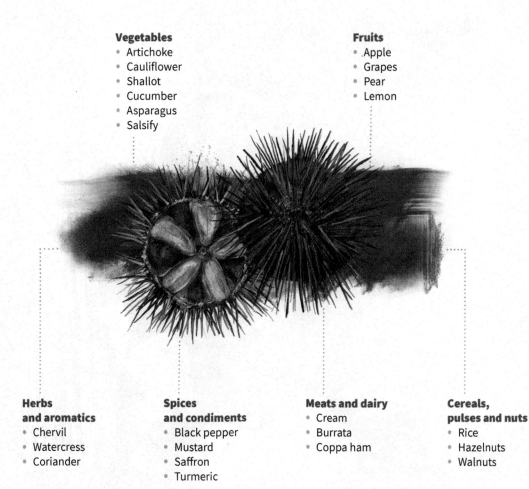

Vegetables
- Artichoke
- Cauliflower
- Shallot
- Cucumber
- Asparagus
- Salsify

Fruits
- Apple
- Grapes
- Pear
- Lemon

Herbs and aromatics
- Chervil
- Watercress
- Coriander

Spices and condiments
- Black pepper
- Mustard
- Saffron
- Turmeric

Meats and dairy
- Cream
- Burrata
- Coppa ham

Cereals, pulses and nuts
- Rice
- Hazelnuts
- Walnuts

Sea urchin risotto with chanterelles and tangerine

Type
main course
Yield
4 servings
Preparation time
15 minutes
Cooking time
25 minutes

This dish is a seafood version of traditional Italian risotto and is flavored with the iodine notes of sea urchin juice, earthy chanterelles and delicate tangerine. The risotto is topped with sea urchin roe and freshly chopped chives.

Utensils needed
- Saucepan
- Fine sieve
- Scissors
- Cutting board
- Fine grater

Ingredients

Aromatic stock
- 1 litre water
- 3 thyme stems
- 1 pinch green fennel seeds
- 2 bay leaves
- 2 garlic cloves

Sea urchin
- 4 sea urchins

Sea urchin risotto
- 2 shallots
- 350 g round risotto rice
- 1 glass of white wine (optional)
- Zest and juice of 2 organic tangerines
- 100 g chanterelles
- 80 g unsalted butter
- 50 g Parmigiano Reggiano
- 1 bunch of chives
- Salt and pepper

Preparation

Aromatic stock
In a saucepan, simmer the water with thyme, fennel seeds, bay leaves and garlic cloves for 10 minutes. Filter the stock through a fine sieve.

Sea urchin
Open the sea urchins with scissors. Use a spoon to carefully remove the roe and gently rinse them under cold water. Filter the sea urchin juice through a fine sieve and reserve for the risotto.

Sea urchin risotto
Peel and finely chop the shallots and clean the chanterelles under cold water. In a saucepan, sauté the shallots and the chanterelles in olive oil. Add the rice and cook on low heat until it turns translucent. Deglaze with white wine, if using, tangerine juice and the sea urchin juice. Cook on low heat, continuously adding the aromatic stock until the rice is al dente. Grate the Parmigiano Reggiano and stir into the risotto along with the butter. Season with salt and pepper.

To serve
Serve the risotto on shallow plates and top with the sea urchin roe and chives.

Nutrition facts

	Per 100 g	Per recipe
Energy	311 kJ/74 kcal	9 895 kJ/2 365 kcal
Protein	2.2 g	57.1 g
Carbohydrate	11.7 g	305.0 g
Fibre	0.6 g	15.4 g
Sugar	0.8 g	21.9 g
Fat	3.4 g	88.7 g
Saturated fats	2.1 g	54.9 g
Sodium	30 mg	790 mg

Seaweed

Gracilaria spp. and *Ulva* spp.

Seaweed in the Mediterranean
and Black Sea region

Scientific name:
Gracilaria spp.

Family:
Gracilariaceae

Approximate average
production volume
(2016–2020): **1 tonne**

The top three producers are
Morocco, Tunisia and **Spain**.

Scientific name:
Ulva spp.

Family:
Ulvaceae

Approximate average
production volume
(2016–2020): **1 tonne**

The top producer is **Spain**.

Production can occur using
ropes and **nets**.

100% of production occurs in
marine water.

Popular among chefs in the
region for their **versatility,
mild flavour** and **health
benefits**.

Source: FAO. 2023. Global aquaculture production quantity (1950–2020).
In: *Fisheries and Aquaculture Division*. Rome. Cited 26 July 2022.
fao.org/fishery/statistics-query/en/aquaculture/aquaculture_quantity

Based on data from the GFCM Information System for
the Promotion of Aquaculture in the Mediterranean (SIPAM)

Gracilaria spp. and *Ulva* spp.

Seaweed

S eaweed is a term commonly used to identify a variety of
marine macroalgae. These macroscopic, multicellular
organisms are classified on the basis of their pigmentation
into three groups: brown, red and green (McHugh, 2003).
Gracilaria spp. is a group of over 100 red seaweeds and a
popular target for aquaculture producers looking to achieve
high yields and source commercially valuable extracts
(Buschmann *et al.*, 2001). On the other hand, *Ulva* spp. are a
diverse group of green seaweeds commonly referred to as
"sea lettuce" that possess compounds highly valuable in
a number of applications. Not only are algae great when
enjoyed on their own, but they also provide extracts that are
useful as functional additives and thickening, gelling and
emulsifying agents for the food and cosmetic industries
(Liao *et al.*, 2021).

Culinary and nutritional value

Despite the high presence and diversity of seaweed species found in the Mediterranean and the Black Sea, their use in the region's traditional cuisines is minor. However, over recent decades, chefs inspired by Asian cuisines have started to highlight the unique sensorial characteristics of algae. *Ulva* spp. are the most used seaweeds in culinary applications across the Mediterranean and the Black Sea and are prized for their slightly iodine flavour and soft texture that is reminiscent of sorrel. Seaweed is a rich source of fibre and antioxidants and is low in calories, making it an ideal ingredient to add to salads, soups or mixed vegetables.

Farming in the region

Seaweeds have been harvested in the Mediterranean and Black Sea region since ancient times; however, marine farming of these macroalgae is a relatively recent concept. In the last 50 years, global seaweed production skyrocketed, reaching almost 35 million tonnes in 2020 (FAO, 2023). Despite this substantial increase in production, seaweed farming is still far from common in many countries of the region, though it has great potential. Contributing to the potential of seaweed is its wide variety of applications in food, feed, biofuel and even specialty biochemicals, all produced alongside many valuable ecosystem services (Neori *et al.*, 2007; Chopin, 2014). In the Mediterranean and the Black Sea, seaweed production largely occurs in Morocco and Tunisia, where farmers produce *Gracilaria* spp. in suspension culture. Elsewhere in the region, similar activities target microalgae and spirulina or work to integrate algae farming into existing operations (GFCM, 2021). Given the global interest in seaweed, as well as the wide range of species and their temperature and latitudinal tolerance, many cultivation methods have been developed (Hanisak and Ryther, 1984). *Ulva* spp. are primarily grown with the help of ropes or nets. Cuttings from existing seaweed are attached to ropes or nets, which are then placed in the open sea, land-based tanks or ponds for the plants to grow to harvestable size (Moreira, 2021). In contrast, *Gracilaria* spp. can be grown by inserting young specimens directly into the sandy floor of a water body and then left to grow to their full size (Capillo *et al.*, 2017). In Morocco and Tunisia, these species can also be cultivated using ropes or tubular nets in lagoons and the open sea.

SELT Marine Group: unlocking the potential of seaweed in the Mediterranean and Black Sea region

Global seaweed production is largely concentrated in Asia, but this exciting sector is now making a mark on the southern Mediterranean region, including through the establishment of the SELT Marine Group seaweed farm in the Bizerte Lagoon.

This Tunisian company was first launched in 1996 as a food technology and processing company for seaweed imported from Asia. Today, after substantial growth, the company produces its own seaweed and, through its processing unit, refines nutritious and valuable seaweed extracts to be used as additives in culinary, cosmetic, pharmaceutical and bioplastic applications, both for local companies and for export. At the core of SELT Marine is a drive to cultivate, process and produce in a way that is environmentally and socially responsible.

Seaweeds are important in the Mediterranean Sea as they provide ecosystem services and contribute to economic and social development in the region. In Tunisia, however, seaweed harvesting has been historically restricted to the Bizerte Lagoon and the Lake of Tunis and has shown untapped potential. It is this fact that inspired the team at SELT Marine to develop their own production.

The farm currently covers 80 ha of the Bizerte Lagoon, exports to 17 countries and is growing quickly. Its success is the result of commitments to continuous innovation, investment in sustainable production methods, ongoing and adaptive training, social development and female empowerment. At the helm of the SELT Marine team, which is composed of over 80 percent women, stands Nadia Selmi, Commercial Manager and an FAO World Food Day food hero.

The Group's principles of environmentally and socially responsible production are evident in both the methods of production and the farm's operations. There are two primary techniques for farming seaweed, particularly *Gracilaria*: "tie tie" and "tubular net". While tie tie is the traditional technique employed in the region and the cheaper of the two, SELT Marine has opted for the more ecologically conscious tubular net technique, which reduces the risk of seeping plastic into the water and damaging the ecosystem. Once harvested, the seaweed is dried naturally with the help of the sun and the wind. Although this practice requires more time, workers, surface and equipment, SELT Marine is

> "Our main objectives are to be independent in terms of raw materials, effectively manage the value chain, and ensure the sustainability of our activities."

committed to its sustainable production strategy and avoids using chemicals. Along with committing to more sustainable production, SELT Marine prioritizes the empowerment of women, employing over 100 women between administration, research and labs and investing in the social development of local communities. It also remains dedicated to education by offering continuous training for staff, adapted based on their roles and goals.

Once seaweed reaches the appropriate size, it is harvested and sent to the transformation unit, where it will be processed into raw powder that can then be used to produce texturizers, gelling agents or food thickeners, in addition to a diverse array of other products. In recent years, the Group has been advancing into the bioplastics sector, using its seaweed extracts in the production of biodegradable and water-soluble bioplastics.

At the moment, SELT Marine is devoting 90 percent of its raw material to the food industry, 5 percent to the pharmaceutical sector and 5 percent to cosmetics, with plans to continue its work in the biomaterials sector.

Additionally, it is expanding its partnerships with businesses around the world and working with research institutions to broaden the scope of useful seaweed-based products.

It has taken a great amount of dedication and effort on the part of the SELT Marine team to reach their current level of success. Initially, the Group faced numerous challenges regarding the exploitation of the lagoon in terms of both administrative constraints and the integration of their operations with other existing activities. In recent years, adaptation to climate change has been the primary focus of SELT Marine, particularly since seaweed growth relies on suitable temperatures.

As SELT Marine continues to expand and diversify, its goals remain the same: continue to secure its autonomy from raw material suppliers by farming its seaweed locally and sustainably, maintain the high quality of its products, and work towards developing innovative and ecological algae-based products, all while remaining committed to environmental and social responsibility.

©GFCM/Daniel Gillet

> "In Tunisia, fish and seafood are typically eaten three or four times a week, but seaweed is not as well known. This sustainable product offers a new taste and texture and can help modernize a traditional dish, such as a Tunisian *tajine*."
>
> **Rihab Nagues, Culinary Student at Institut Paul Bocuse**

©SELT Marine Group

©SELT Marine Group

Traditions and recipes from the region

S eaweed is fast becoming an ingredient of interest for chefs due to its diversity of flavour, colour and texture. Inspired by Asian and northern European cuisines, various seaweed species are used in the preparation of pastas, breads, vegetables and fish. Found fresh, brined, dried or reduced into a powder, seaweed offers a versatility of use and can be easily incorporated into a number of recipes.

Seaweed *focaccia*

Ulva spp. are used in various bread recipes, offering an iodine flavour and giving the bread a unique visual identity. *Ulva* spp. pair ideally with the olive oil used in the preparation of Italian *focaccia*. Added finely minced or in dried flakes at the end of the fermentation process, the seaweed will perfume the dough during the baking process. The quantity of salt in the dough should be reduced, as the seaweed will add salinity. The *focaccia* can be topped with grilled vegetables or with slices of smoked fish.

Seaweed risotto

Seaweeds such as *Ulva* spp. can be used in the preparation of various rice and pasta recipes, including risottos, to develop an iodine flavour. To be included in a risotto, *Ulva* spp. seaweeds should be rinsed and torn apart in fine strips. After the strips are poached for several minutes in a vegetable stock to soften their texture, the stock is then used to flavour the risotto. The strips are quickly sauteed with olive oil and added to the risotto just before it is served.

Seaweed tartare

Ulva spp. tartare is prepared by soaking, rinsing and finely mincing seaweed, which is then seasoned with shallots, onions, capers, olives, garlic, spices and fresh herbs to develop its aromatic profile. *Ulva* spp. tartare is served on toast, used as a condiment to accompany a dish or as a filling for baked vegetables.

Chef's tips

How to prepare *Ulva* spp. seaweed?

Seaweed is often found brined or dehydrated. In both cases, it is essential to soak the seaweed in clear water for at least one hour in order to remove the excess salt. It can also be blanched for 30 seconds and cooled down immediately in ice water to soften its texture and facilitate its addition into dishes.

How to cook *Ulva* spp. seaweed?

Rehydrated seaweed offers a diversity of applications. Finely chopped, it can flavour a sauce or be incorporated into a bread dough. Sliced into fine strips, *Ulva* spp. seaweed can be eaten raw in a salad or pickled to add to warm toast. Torn into larger strips, it can be sauteed with vegetables, poached in a broth or used to wrap fish fillets in order to protect them from the heat when roasted.

How to season *Ulva* spp. seaweed?

Seaweed is often used as a seasoning to bring iodine notes to the preparation of fishes, sauces or even flavoured butter. However, seaweed's interesting texture can also be highlighted in spreads, in the preparation of soups or sauteed with vegetables. The distinct aromatic profile of Ulva spp. seaweed pairs well with fresh herbs, garlic and capers.

Here are the best ingredients to pair with *Ulva* spp. seaweed for unique Mediterranean and Black Sea-inspired dishes

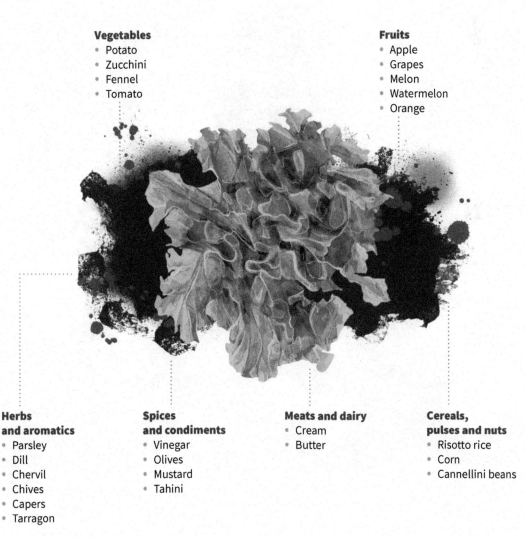

Vegetables
- Potato
- Zucchini
- Fennel
- Tomato

Fruits
- Apple
- Grapes
- Melon
- Watermelon
- Orange

Herbs and aromatics
- Parsley
- Dill
- Chervil
- Chives
- Capers
- Tarragon

Spices and condiments
- Vinegar
- Olives
- Mustard
- Tahini

Meats and dairy
- Cream
- Butter

Cereals, pulses and nuts
- Risotto rice
- Corn
- Cannellini beans

Seaweed tajine with pistachios and dates

Type
main course
Yield
4 servings
Preparation time
15 minutes
Cooking time
20 minutes

Tajine is emblematic of Maghrebian cuisine. While usually prepared with meat or fish, this version is a vegetarian option flavoured with seaweed. The iodine aromas are paired with roasted vegetables perfumed with spices, sweet apricots, dates and fresh notes from coriander leaves.

Utensils needed
- Cutting board
- Non-stick pan
- Vegetable peeler
- Saucepot
- Saucepan
- Plates
- Fine grater

Ingredients

Seaweed tajine
- 100 g seaweed
- 1 white onion
- 2 garlic cloves
- 3 tbsp extra virgin olive oil
- 3 turnips
- 2 carrots
- 2 zucchini
- 4 potatoes
- 4 preserved lemons
- 1 tbsp ground ginger
- 1 tbsp ground turmeric
- 1 tbsp ground coriander
- 1 tbsp ground cinnamon
- Salt and pepper

Seaweed semolina
- 300 g fine semolina
- 2 tbsp seaweed
- 300 ml water
- 2 tbsp extra virgin olive oil
- Salt and pepper

Topping
- 4 dates
- 4 dried apricots
- 1 handful of almonds
- 1 handful of pistachios
- ½ bunch of fresh coriander

Preparation

Seaweed tajine
Rinse the seaweed under cold water. Peel and finely chop the onions and garlic. Peel and chop the carrots, turnips, potatoes and zucchini. Cut the preserved lemons into four pieces. In a pot, sauté the onion and garlic in olive oil, then add the vegetables, preserved lemon pieces, seaweed and spices and cook, covered, over low heat for 20 minutes, until the vegetables are cooked through.

Seaweed semolina
Bring the water to boil, add the semolina and seaweed and cover with a cloth. Let the semolina hydrate for 5 minutes, then mix using a fork. Add olive oil and season with salt and pepper.

To serve
Pit and finely slice the dates. Finely slice the apricots. Chop the almonds and pistachios. Finely chop the coriander. Scoop the seaweed semolina onto a shallow plate, followed by the tajine vegetables. Top with the dates, apricots, almonds, pistachios and coriander.

Nutrition facts

	Per 100 g	Per recipe
Energy	392 kJ/94 kcal	12 569 kJ/3 004 kcal
Protein	2.3 g	73.6 g
Carbohydrate	13.1 g	419.0 g
Fibre	2.3 g	72.9 g
Sugar	3.5 g	111.0 g
Fat	2.9 g	93.3 g
Saturated fat	0.4 g	11.3 g
Sodium	40 mg	1 279 mg

Tilapia

Oreochromis spp.

**Tilapia in the Mediterranean
and Black Sea region**

Scientific name:
***Oreochromis* spp.**

Family:
Cichlidae

Approximate average
production volume
(2016–2020):
1 000 000 tonnes

The top three producers are
Egypt, **Israel** and the **Syrian
Arab Republic**.

88% of production occurs in
brackish water and
12% in **freshwater**.

Healthy, **affordable** and
mild in flavour, it is a great
option for family meals.

Source: FAO. 2023. Global aquaculture production quantity (1950–2020).
In: *Fisheries and Aquaculture Division*. Rome. Cited 26 July 2022.
fao.org/fishery/statistics-query/en/aquaculture/aquaculture_quantity

Based on data from the GFCM Information System for
the Promotion of Aquaculture in the Mediterranean (SIPAM)

Oreochromis *spp.*

Tilapia

The farming of tilapia (*Oreochromis* spp.) dates back over 4 000 years (Gupta, 2004). Today, the increasing commercialization and growth of the aquaculture industry makes it one of the most important farmed species groups of the twenty-first century (Shelton, 2002). This tropical fish thrives in warm waters and its omnivorous behaviour allows for fast and sustainable growth through the filtering and grazing of phytoplankton, zooplankton, mucus and detritus. The most common tilapia species in the Mediterranean and Black Sea region is Nile tilapia (*Oreochromis niloticus*).

Culinary and nutritional value

Tilapia is a widely consumed fish in southeastern Mediterranean countries, especially in Egypt. Its affordably priced, mild, pinkish flesh makes it a prime choice for daily family meals. It is an excellent source of vitamin B and essential fatty acids and can be prepared in various ways, including grilled on an open fire, breaded and pan-fried, or served in a broth. The flesh becomes dry when overcooked, hence quick cooking processes at high temperatures are preferred.

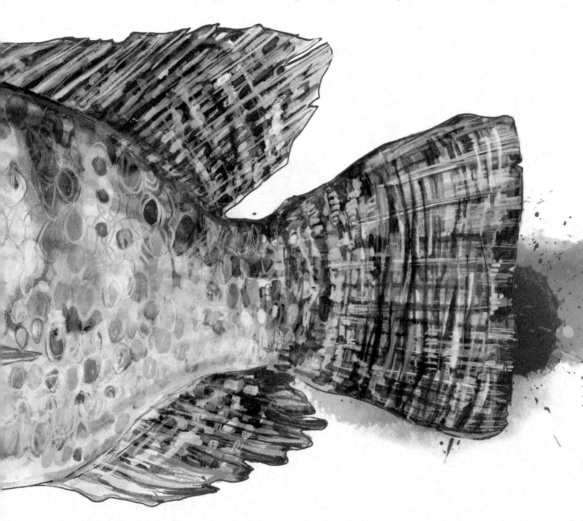

Farming in the region

The first recorded instance of controlled culture of tilapia occurred in Kenya in 1924 and soon spread throughout Africa before expanding to Asia and the Americas in the following decades (Gupta, 2004). Now, tilapia is farmed in 85 countries worldwide, including in the Mediterranean and the Black Sea. While several species of tilapia are farmed commercially in the region, Nile tilapia is the most prominent. The production process begins with the spawning of the species, which can occur at any time of year as long as the water temperature reaches 24°C. Over the next 10 to 15 days, the eggs are incubated and begin to hatch. The young tilapia are then nursed until they reach an average size of 30–40 g, a process that takes approximately 2–3 months (FAO, 2022f). For the final phase of tilapia production, producers can opt to use ponds, floating cages, tanks, raceways, or recirculating systems depending on their needs. When the tilapia reach the desired commercial size, producers using ponds conduct a complete harvest and those using tanks, raceways and recirculating systems conduct a partial harvest to maximize production. The future for tilapia aquaculture looks very promising as this fish is an ideal candidate for integration with crop and animal farming (Wang, M. and Lu, M., 2015).

Mohamed Mahmoud Kord: a pharaonic tradition brought into the modern age

The fertile Nile river valley paints a dense ribbon of luscious farmland through the desert landscape of Egypt. Above Cairo, the waterway frays into many branches, leaving a fan-shaped green stamp on the northern coast. Here, Mohamed Mahmoud Kord's father worked as a fisher in the brackish waters of Lake Burullus, separated from the Mediterranean Sea by a thin beach. In 1980, he decided to try his luck on dry land, purchasing four hectares of farmland nearby. The plot was too close to the sea, however, and even amongst the surrounding abundance, the salty soil produced few crops. He turned his attention back to fish, digging two shallow ponds on his land to raise his own.

Though ancient tomb illustrations testify to the aquaculture of Nile tilapia in man-made ponds as early as 4 000 years ago, the art of farming the species in Egypt had been long forgotten by 1980. Tilapia's ability to tolerate exceptionally muddy, low-oxygen conditions was assumed to reflect poorly on the quality of its flesh and it had developed a reputation as a "trash fish." Gone were the days of tilapia amulets and hieroglyphs; there was

little interest for farm-raised tilapia in Egypt. Instead, Kord's father, with the help of young Mohamed, struggled to raise carp and mullet, slower-growing and more expensive fish. It wasn't until tilapia returned to Egyptian markets in the 1990s, riding a wave of expert and public re-evaluation and new aquaculture technologies, that the Kords' farm found its stride.

Between 1990 and 1995, Nile tilapia hatcheries and feed factories were opened by government ministries and private companies alike. In the face of mounting fishing pressure on native marine stocks, tilapia's fast growth rates, simple dietary needs and success in crowded environments encouraged aquaculture producers to bet on the fish. The Kords purchased fry from the new hatcheries and feed from the factories and introduced tilapia to their mullet ponds. Their annual production quickly jumped from about 800 kg per hectare to 2 500 kg. In 2000, the Kords more than quadrupled the size of their farm, purchasing six additional hectares of land and renting another ten.

As Mohamed Kord took on more responsibility in his family's business, he kept abreast of the newest developments in fish farming techniques. While pursuing studies at Alexandria University's Department of Agriculture, he applied practices picked up from the classroom, as well as his own innovations, to boost production. He switched to the use of extruded feed, which is compressed, coated and dried in a factory setting, making it a cheap and efficient product perfect for farmers. Later,

©Mohamed Mahmoud Kord

©GFCM/Daniel Gillet

he installed a paddle wheel to maintain the optimum level of dissolved oxygen in his ponds and adopted advanced post-harvest and transportation procedures to ensure the safety and freshness of his products.

Kord also travelled independently to enrich his knowledge of fish farming, from China to Egypt's very own WorldFish research centre in Abbasa. Over the last decade, Kord has transitioned to the other side of the lectern, offering his personal expertise to aspiring Egyptian fish farmers. He was selected by WorldFish as one of twenty certified trainers to teach best aquaculture methods to thousands of Egyptian fish farmers.

Kord belongs to the first generation of Egyptian tilapia farmers since the days of the pharaohs, and he and his colleagues are advocating for improvements in the industry to keep successfully expanding. For example, tilapia die-off over the summer from disease is common. Additionally, the fish is marketed whole in Egypt, but building factories to produce fillets and other value-added products could open the door to greater exports. Given the rapid growth of the sector so far, it seems unlikely that these concerns will go unresolved for long.

"In Egypt, fish consumption depends on the area you're in; closer to the sea, we eat more fish because it's fresh. One of the most popular fish in Egypt is tilapia. It can be prepared by marinating it in a mixture of garlic, hot chili, peppers and spices before baking or frying and serving it alongside rice and fresh salads."

Joude El Shennawy,
Culinary Student at Institut Paul Bocuse

"My goal is to further the development of aquaculture in my region and to produce a distinct fish with good specifications for the Egyptian market and international export."

Traditions and recipes from the region

Tilapia has been present in Egyptian culture for over 4 000 years. An ancient Egyptian recipe recommends that tilapia be cut into pieces leaving the skin and bones intact, then cooked in a barley stew and served with shallots. In pharaonic times, it was also cooked directly over a flame on a spit. Today, it is regularly prepared grilled on a barbecue or fried whole and served in affordable restaurants.

©GFCM/Dominique Bourdenet

Egypt / **Tilapia soup with barley**

In Egypt, Tilapia is stewed in an oven dish with vegetables and spices. The fish is gutted and scaled before being cut into medium-sized slices with the bones intact. The slices are arranged in a baking dish along with garlic, chili powder, celery leaves and coriander, and sprinkled with various spices. Vegetables cut into sticks and more of the spice mixture are added and the dish is finished with pieces of butter and the freshly squeezed juice of an orange. It is then covered with aluminum foil and cooked for one hour.

©GFCM/Ziad Samaha

Lebanon / *Kebbet samak*

Kebbet is one of the most representative dishes of Lebanese cuisine. Usually prepared with minced lamb and bulgur, a fish version can be found in the region of Beirut. White fish such as tilapia is filleted and ground with bulgur, sauteed onions, coriander and parsley. The preparation is spiced with orange zest, cumin, coriander, cinnamon, turmeric and white pepper, then shaped into the traditional kebbet form. It is then deep-fried and served with houmous or tahini sauce.

©GFCM/Galatea Media

Syrian Arab Republic / *Sayadiyah*

In this traditional recipe from the Syrian coast, fillets of white fish such as tilapia are baked in an earthen pot with a lemon, coriander seed, garlic and olive oil marinade. The fish is served with rice perfumed with a spice mix called baharat, which is principally composed of cumin, caraway and coriander and garnished with caramelized onions. Across the region and among families, the spice mix can vary by adding, for example, cinnamon, turmeric or safflower.

Chef's tips

How to prepare tilapia?

Because of its affordable price, tilapia can be used as the main ingredient in the preparation of fish balls, fillings or spreads. The fillets are commonly found fresh or frozen without skin and fishbones. However, to prevent any risk, check that there are no fishbones left in the fillet before going through the effort of preparing the fish.

How to cook tilapia?

Tilapia flesh can easily release water during the cooking process and become dry. To facilitate the preparation of fillets, they can be sliced finely and sauteed quickly with olive oil in a very hot non-stick pan. Tilapia flesh can also be roasted and shredded and then served cold as a spread with olives, or warm as a fish gratin.

How to season tilapia?

Compared to high-value fishes like turbot or seabass, tilapia is more adaptable and can be transformed in a variety of recipes, rather than simply pan-fried or roasted. The mild flesh can be paired with a variety of ingredients, from a classical breaded version with parsley and garlic to a spicy fish spread with grilled red bell peppers and capers.

Here are the best ingredients to pair with tilapia for unique Mediterranean and Black Sea-inspired dishes

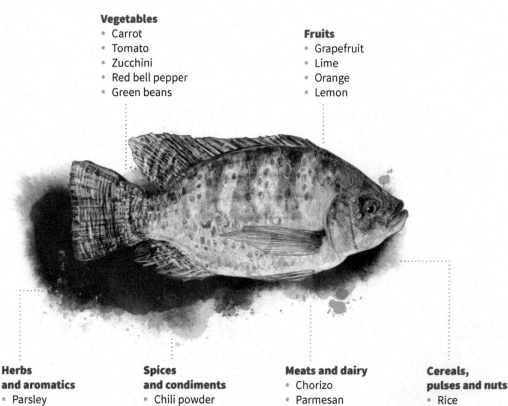

Vegetables
- Carrot
- Tomato
- Zucchini
- Red bell pepper
- Green beans

Fruits
- Grapefruit
- Lime
- Orange
- Lemon

Herbs and aromatics
- Parsley
- Rosemary
- Sage
- Coriander
- Tarragon

Spices and condiments
- Chili powder
- Smoked paprika
- Turmeric
- Dried garlic
- Olives
- Nigella

Meats and dairy
- Chorizo
- Parmesan

Cereals, pulses and nuts
- Rice
- Bulgur
- Faba beans
- Almonds

Tilapia sandwich with confit red pepper, grilled zucchini and a dill and tahini dressing

Type
main course
Yield
4 servings
Preparation time
10 minutes
Cooking time
40 minutes

This dish, inspired by Turkish street food, is prepared with thinly-sliced spiced tilapia and grilled vegetables served inside pita bread as a kebab. It is paired with a distinct toasted sesame and fresh dill sauce.

Ingredients

Dill pita bread
- 5 dill stems
- 300 g flour, type 55
- 25 g fresh yeast
- 100 ml whole milk
- 75 ml water
- 1 pinch garlic powder
- 2 pinches poppy seeds

Dill and tahini sauce
- 5 tbsp tahini
- 5 tbsp water
- Juice of ½ lemon
- 1/4 bunch of dill
- 2 garlic cloves
- 1 tbsp extra virgin olive oil
- Salt and pepper

Confit red bell pepper
- 2 red bell peppers
- Salt and pepper

Pan-fried zucchini
- 2 zucchini
- 2 tbsp extra virgin olive oil
- Salt and pepper

Marinated tilapia
- 4 tilapia fillets
- 2 tbsp extra virgin olive oil
- 1 tbsp ground ginger
- 1 tbsp ground cumin
- 1 tbsp ground coriander

Utensils needed
- Cutting board
- Mixing bowl
- Towel
- Baking tray
- Greaseproof paper
- Blender
- Non-stick pan
- Fish bone tweezers

Preparation

Dill pita bread
Finely chop the dill stems and mix with the flour, yeast, milk, water, garlic powder and poppy seeds. Knead for 5 minutes, then cover with a towel and let rest for 30 minutes at room temperature. Divide the dough into four, shape each piece into a ball and roll to 1 cm thickness. Place the four pieces of dough on a baking tray lined with greaseproof paper and bake in an oven preheated to 180 °C for 12 minutes.

Dill and tahini sauce
Chop the garlic and the dill. In a blender, combine the tahini, water, lemon juice, dill, garlic and olive oil until smooth. Season with salt and pepper.

Confit red bell pepper
On a baking tray lined with greaseproof paper, roast the whole red peppers in an oven preheated to 180 °C for 20 minutes. Remove the skin and seeds of the bell peppers and cut in strips. Season with salt and pepper.

Pan-fried zucchini
Cut thin slices of zucchini lengthwise. In a non-stick pan, pan-fry the zucchini slices in olive oil over high heat until golden-brown. Season with salt and pepper.

Marinated tilapia
Remove the fishbones and skin from the tilapia fillets. Thinly slice the fillets. In a bowl, mix the olive oil and spices, add the tilapia slices and allow to marinate for 10 minutes. In a non-stick pan, pan-fry the tilapia pieces over medium heat for 2 minutes.

To serve
Cut the pita bread in half and garnish with the sauce, confit bell pepper, zucchini and tilapia.

Nutrition facts

	Per 100 g	Per recipe
Energy	415 kJ/99 kcal	11 962 kJ/2 859 kcal
Protein	5.5 g	157.0 g
Carbohydrate	9.9 g	385.0 g
Fibre	1.4 g	41.3 g
Sugar	1.3 g	37.3 g
Fat	3.8 g	109.0 g
Saturated fat	0.6 g	17.9 g
Sodium	41 mg	1 174 mg

Turbot

Scophthalmus maximus

Turbot in the Mediterranean
and Black Sea region

Scientific name:
Scophthalmus maximus

Family:
Scophthalmidae

Approximate average
production volume
(2016–2020):
7 800 tonnes

The top producer is **Spain**.

99% of production
occurs in **brackish water**
1% in **marine water**.

Low in fat and **high in protein**,
it is a great option for healthy
recipes.

Source: FAO. 2023. Global aquaculture production quantity (1950–2020).
In: *Fisheries and Aquaculture Division*. Rome. Cited 26 July 2022.
fao.org/fishery/statistics-query/en/aquaculture/aquaculture_quantity

Based on data from the GFCM Information System for
the Promotion of Aquaculture in the Mediterranean (SIPAM).

Scophthalmus maximus

Turbot

Turbot (*Scophthalmus maximus*) is a member of the Pleuronectiformes order, commonly called flatfish. This species is found in the Mediterranean and the Black Sea as well as in the Northeast Atlantic, including the Baltic Sea (Bauchot, 1987). Turbot has proven to be important not only for capture fisheries, but also for aquaculture in the Mediterranean and the Black Sea, as it is appreciated for its mild flavour and health benefits. Given these desirable characteristics, production has rapidly increased, particularly in Spain, to meet demand for this highly prized fish.

Culinary and nutritional value

Known as the "prince of the seas", turbot is considered to be one of the finest, and therefore highest-priced, fish in the Mediterranean and the Black Sea. The four fillets in each of these flatfish have a creamy white colour, a delicate perfumed taste and a firm texture. Turbot can be prepared whole, in fillets or in pieces with the bones and skin intact, or it can be braised, grilled on an open fire or delicately poached. Its flesh is low in fat and high in protein, making it an ideal choice for healthy recipes.

Farming in the region

Turbot has always been a highly appreciated fishery product and for this reason, the aquaculture sector worked tirelessly to convert this species into a farming possibility. Turbot production began in Scotland in the 1970s and by the 1980s had expanded to continental Europe (FAO, 2022g). An initial small number of farms in Spain quickly multiplied following improvements in juvenile rearing technologies, leading to a boom in production. Later, due to high production costs and an unconsolidated commercial market, many producers were forced to close (FAO, 2022g). Today,

production in the Mediterranean and Black Sea region has expanded to numerous countries, including Italy, Portugal and Türkiye but remains concentrated in Spain, which boasted an approximate production volume of 7 000 tonnes in 2020 (FAO, 2023).

The production cycle for turbot begins by obtaining fish larvae from adult specimens: broodstock are held in tanks and spawning is conducted by manually stripping the fish. Once the eggs hatch, the turbot larvae are nursed in tanks and fed rotifers, artemia and phytoplankton before being weaned to dry pellets when

they reach 5–10 g. During this life stage, each larva undergoes metamorphosis, moving one eye to the other side of its body and becoming distinctively flat. After four to six months, the juveniles reach 80–100 g, and producers start the on growing phase. During this phase, turbot are reared in flat bottomed cages or on shore tanks (by far the more common method of the two). It takes approximately 18 to 20 months for turbot to reach a commercial size of 1.5–2 kg, during which time they are given dry feed and kept at a temperature of 14–18 °C. Once the turbot reach commercial size, producers harvest the fish by hand (FAO, 2023; Çiftci *et al.*, 2002).

Central Fisheries Research Institute: developing a model for the private sector to follow

In Türkiye, the government has taken the lead on aquaculture. The Central Fisheries Research Institute (SUMAE) in Trabzon, with its state-of-the-art laboratories, indoor fish tanks, outdoor ponds and gene bank, is the site of frequent groundbreaking advances in farmed fish production, gene technologies and natural stock enhancement. The Institute plays a crucial role in supporting an aquaculture industry that boasts an estimated export target for 2023 of USD 1 billion. Much of SUMAE's aquaculture research and resources are directed toward turbot, one of the Black Sea's characteristic species and an economically vital fish, whose wild stocks are at risk due to overexploitation.

The diverse genetic traits and life histories of the over 40 000 turbot individuals born at

SUMAE each year testify to the wide range of innovative experiments undertaken at the Institute. In addition to hatching turbot larvae during the traditional spawning season in May, SUMAE scientists are able to produce turbot year-round through the photoperiod manipulation technique. This process involves subjecting fish to different light regimes – exposing them to more or less light over the course of a day than would be encountered under natural conditions – in order to accelerate or delay key developmental stages in the turbot's life history.

The Institute's gene bank backstops all of these procedures. Liquid nitrogen freezers maintained at temperatures of -196 °C host high-quality sperm from up to 35 different male turbot, either raised on SUMAE grounds or sourced from wild populations. The extremely low temperatures prevent the formation of ice crystals that occur during normal freezing, which can damage organic material. If breeding efforts are met with unexpected problems one season, especially in the collection of sperm, SUMAE researchers turn to the gene bank and its extensive reserves.

Turbot production at SUMAE began in the late 1990s with the help of the Japan International Cooperation Agency (JICA). Experts from Japan, where several of turbot's flatfish cousins, such as flounder and halibut, have been successfully farm-raised for decades, spent time at SUMAE teaching techniques in flatfish cultivation, while Turkish scientists traveled the opposite direction to learn on-site in Japan.

Most of the turbot production at SUMAE over the last 25 years has been for research purposes, usually led by in-house scientific teams or university groups. However, according to SUMAE, the objective of this research is "to create an exact protocol for the private sector to produce year-round successful production." Indeed, cooperation with private companies has already been well established through a variety of ways: surplus production of turbot larvae often goes to turbot farmers in the region, representatives from the private sector visit SUMAE to gain expertise and knowledge, and companies even help in financing research projects at the Institute. "From the perspective of successfully applied techniques, the achievements of this species' culture here are crucial for the private sector to start producing the species," SUMAE affirms.

Beyond the private sector, SUMAE acts as a specialized hub to share knowledge with all aquaculture stakeholders. As one of two GFCM Aquaculture Demonstration Centres in the Black Sea, SUMAE promotes sustainable aquaculture in the region by sharing knowledge via demonstrative trainings, promoting technical cooperation and increasing capacity.

Another area in which SUMAE's turbot production actively supports socioeconomic growth is capture fisheries, through restocking programmes. Approximately 10 000 of the juveniles produced at the Institute each year are destined for life in the Black Sea, and while these fish are tagged and monitored for scientific studies to gauge the success of stock enhancement efforts, the private sector once again benefits. Turbot's natural habit is to disperse

"Fish is very popular in Italy, with every region favouring their preferred species and method of preparation. Turbot is a popular choice and its fillets can be served pan-fried with olive oil and garlic and accompanied by potaoes, tomatoes and olives cooked in a turbot-bone fumet."

Luca Violi, Culinary Student at Institut Paul Bocuse

minimally from release sites, so the fish are often found within the same general area when they reach minimum catch size. Local fishers greatly appreciate the influx of new turbot to their waters, which they know may lead to increased landings and enhanced aquatic food production without exceeding the natural productivity of an important, but overexploited, commercial stock.

Farmed turbot production is a challenging enterprise that requires large economic investments and technical know-how. Dedicated research by the best-qualified scientists in the country has made Türkiye one of the few places in the world where commercial scale turbot cultivation is underway. As the industry finds its feet in the private sector, SUMAE has itself made plans to sell market-size turbot raised in its outdoor circular concrete ponds to local consumers at a reasonable price.

"The Institute's turbot production actively supports socioeconomic growth through restocking programmes."

Traditions
and recipes from the region

The earliest reference to this fish comes from a Roman satirical poem dating from the end of the first century CE, which suggests that it was a very popular dish in the Roman Empire. Today, turbot is an expensive fish, highly prized for its delicate taste. It can be cooked in a variety of ways: in the oven, grilled or poached. To bring out the best of its fine-tasting flesh, avoiding lengthy cooking processes is recommended.

©Florian Olivo

France / *Turbot à la Grenobloise*

This traditional French recipe is prepared with highly appreciated fishes such as turbot. The turbot is scaled and gutted, then rinsed with clear water and dried with a clean towel. The fish is then either prepared whole, to be served at the centre of the table, or cut into segments. To prevent the fish from sticking, it is coated with flour before being pan-fried with butter until the skin is crispy and golden-brown. It is served with croutons, capers, chopped parsley and lemon slices.

Georgia / **Roasted turbot with eggplants and pomegranate sauce**

The traditional Georgian method of preparing fish from the Black Sea, such as turbot, is to scale and gut the fish, before rubbing it with salt, crushed garlic and olive oil and roasting it in an oven until the inner flesh is done and the skin turns golden-brown. It is served with slices of pan-fried eggplants in olive oil and a sauce prepared with pomegranate juice reduced to a syrup and seasoned with onions, minced garlic, basil and cilantro.

©GFCM/Dominique Bourdenet

©GFCM/Claudia Amico

Türkiye / *Balık ekmek*

This fish sandwich prepared with the fresh catch of the day is Istanbul's iconic street food. Commonly featuring mackerel, a rarer version of this sandwich made with turbot and found in higher end-restaurant stands out as one of the best recipes. The fish is grilled on an open fire and served between two slices of fresh bread accompanied by slices of white onions, salad leaves and slices of tomato. A slice of lemon is served with the sandwich to squeeze on top at the last moment.

Chef's tips

How to prepare turbot?

Turbot is a flatfish consisting of four fillets, two bigger ones on the darker side and two smaller ones on the lighter side. To remove the fillets, cut from the tail to the head following the white line representing the backbone of the fish. Then, using a filleting knife, carefully detach the fillets from the fishbones. Keep the fish's bones and skin, as they can be transformed into an excellent fish fumet.

How to cook turbot?

Turbot flesh is sensitive to long heat treatment and easily becomes dry when overcooked. When braising, grilling, roasting or even cooking with a spit on an open fire, prefer using a whole turbot or cutting it into segments while keeping the skin and bones intact. These alternatives will protect the delicate flesh while cooking.

How to season turbot?

As turbot is a high-value fish, it is imperative to select the right ingredient pairings to preserve its sensorial qualities. The fish simply pan-fried in olive oil with a pinch of salt is already a delicacy. Ingredient pairings should complement its flesh with subtle and noble flavours, like those of chanterelles or green asparagus.

Here are the best ingredients to pair with turbot for unique Mediterranean and Black Sea-inspired dishes

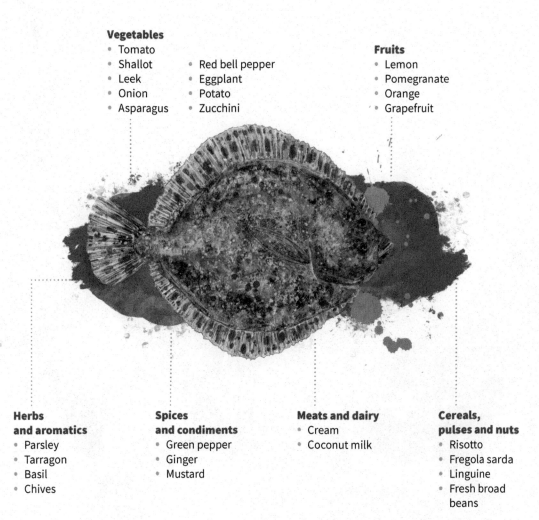

Vegetables
- Tomato
- Shallot
- Leek
- Onion
- Asparagus
- Red bell pepper
- Eggplant
- Potato
- Zucchini

Fruits
- Lemon
- Pomegranate
- Orange
- Grapefruit

Herbs and aromatics
- Parsley
- Tarragon
- Basil
- Chives

Spices and condiments
- Green pepper
- Ginger
- Mustard

Meats and dairy
- Cream
- Coconut milk

Cereals, pulses and nuts
- Risotto
- Fregola sarda
- Linguine
- Fresh broad beans

Pan-fried turbot and eggplant tian with arugula and basil pesto

Type
main course
Yield
4 servings
Preparation time
20 minutes
Cooking time
30 minutes

In this recipe, delicate pan-fried turbot is paired with an eggplant tian inspired by the traditional recipe from the south of France and a twist on Italian basil pesto.

Ingredients

Utensils needed
- Cutting board
- Baking tray
- Greaseproof paper
- Baking tin
- Fine grater
- Blender
- Scissors
- Fishbone tweezers
- Non-stick pan
- Plates

Eggplant tian
- 4 eggplants
- 1 tsp extra virgin olive oil
- Salt and pepper
- 2 sprigs of fresh thyme
- 2 garlic cloves
- 1 tsp pine nuts
- ½ jar of anchovies
- ½ jar of capers
- 2 tbsp taggiasca olives

Arugula and basil pesto
- 2 garlic cloves
- 2 tbsp extra virgin olive oil
- 100 g Parmigiano Reggiano
- 1 bunch of fresh basil
- 2 handfuls of pine nuts
- 1 handful of arugula
- Salt and pepper

Pan-fried turbot
- 1 small turbot
- 1 tbsp extra virgin olive oil
- Fine salt

Preparation

Eggplant tian
Wash and thinly slice the eggplants. Place them on a baking tray lined with greaseproof paper, sprinkle with olive oil and season with salt and pepper. Top with sprigs of thyme and garlic cloves and bake in an oven preheated to 180 °C for 15 minutes. Form the tian by vertically layering the eggplant slices in a baking tin. Top with pine nuts, anchovies, capers and taggiasca olives and bake for 10 minutes.

Arugula and basil pesto
Peel the garlic cloves and blend together with the olive oil, Parmigiano Reggiano, basil, pine nuts and arugula until smooth. Season with salt and pepper.

Pan-fried turbot
Remove the scales, the fins and the gills of the turbot. Rinse the fish under cold water. Remove the head and tail and cut the turbot into four pieces. Season with fine salt. In a non-stick pan, pan-fry the turbot in olive oil over medium heat for 3 minutes on each side, until the skin is golden and crispy.

To serve
Place each piece of turbot on a flat plate with the eggplant tian and the arugula and basil pesto on the side.

Nutrition facts

	Per 100 g	Per recipe
Energy	418 kJ/100 kcal	7 100 kJ/1 697 kcal
Protein	5.5 g	93.0 g
Carbohydrate	2.7 g	46.1 g
Fibre	2.3 g	39.3 g
Sugar	2.1 g	35.5 g
Fat	6.9 g	117.0 g
Saturated fat	1.8 g	29.8 g
Sodium	198 mg	3 346 mg

European seabass

Dicentrarchus labrax

European seabass in the Mediterranean and Black Sea region

Scientific name:
Dicentrarchus labrax

Family:
Moronidae

Approximate average
production volume
(2016–2020):
232 500 tonnes

The top three producers are
Türkiye, **Greece** and **Egypt**.

86% of production
occurs in **marine water** and
14% in **brackish water**.

An iconic fish, typically
served in higher-end
restaurants, **now becoming
more affordable thanks to
aquaculture**.

Source: FAO. 2023. Global aquaculture production quantity (1950–2020).
In: *Fisheries and Aquaculture Division*. Rome. Cited 26 July 2022.
fao.org/fishery/statistics-query/en/aquaculture/aquaculture_quantity

Based on data from the GFCM Information System for
the Promotion of Aquaculture in the Mediterranean (SIPAM)

Dicentrarchus labrax

European seabass

European seabass (*Dicentrarchus labrax*), a member of the Moronidae family, was the first non-salmonid marine species to be commercially cultured in Europe and has become one of the most important commercial fish cultured in the Mediterranean and Black Sea region, reaching an approximate production volume of 270 500 tonnes in 2020 (FAO, 2023). This predator species can inhabit coastal areas, estuaries and other waters, all with a wide range of salinities (FAO, 2022h).

Culinary and nutritional value

European seabass is an iconic fish served in higher-end restaurants along the coasts of the Mediterranean Sea. Its flesh is white and tender, rich in vitamin B_{12} and omega-3 fatty acids. The delicate flavours developed through grilling, pan-frying or roasting offer a premium culinary experience. Aquaculture has introduced affordable seabass to the market, making it a convenient choice for festive meals.

Farming in the region

Historically, farming of European seabass was tied to salt production in coastal evaporation pans and marshes, with the salt harvest occurring during the summer and autumn and fish culture during the winter and spring. By the 1970s, however, innovative techniques developed in France and Italy allowed producers in most Mediterranean countries to begin mass-producing juvenile seabass, resulting in harvests of hundreds of thousands of larvae (FAO, 2022h). Today, the largest producers of European seabass in the region are Türkiye, Spain, Greece and Egypt, which together produced approximately 245 000 tonnes in 2020 (FAO, 2023). The majority of this volume – an impressive 83 percent – comes from sea cage farming, although the species is traditionally farmed in seawater ponds and lagoons. Cage systems rely on the natural exchange and qualities of marine water and are generally used in intensive productions involving seabass juveniles obtained from commercial hatcheries rather than from wild stocks. Typically, cages are made of a net suspended below a floating circular or square frame. These structures can be anchored near the coast, accessible by a landing, or located offshore to be reached only by boat. The ongrowing phase lasts approximately 18 to 24 months until the fish reach 400 g to 450 g – the common market size – and therefore are ready to be harvested (FAO, 2022h).

©Kefalonia Fisheries

Kefalonia Fisheries:
a family's labour of love
goes international

Along the Paliki Peninsula's eastern coast, running parallel to the cliffy mainland beaches of the Greek island of Kefalonia, verdant pastures, vineyards and olive groves meet the brilliant blue Gulf of Argostoli. For over a century, the Geroulanos family has been sustainably stewarding the natural resources on both sides of the shoreline as fishers and farmers. Today, Lara Barazi-Geroulanou is the CEO of Kefalonia Fisheries, the very first Greek fish farm to produce seabass in this deep inlet of the Ionian Sea.

Since Marinos Geroulanos founded the company in 1981, raising its fish has been a labour of love, drawing upon the best of both traditions of farming and fishing. Now in the hands of the second generation of the family, Kefalonia Fisheries is distinguished by close-knit corporate culture across its management, production and sales divisions, a large number of whom are women blazing their own trail in a traditionally male-dominated industry. Dedicated to exalting the natural

flavours of the Mediterranean while leaving a minimal impact on the local environment, the team has worked hard to earn a number of certifications for its sustainable and responsible practices, including an organic certification for its products awarded by the European Union and a label of recognition from the Aquaculture Stewardship Council (ASC).

Kefalonia Fisheries raises its stocks of European seabass – and gilthead seabream (*Sparus aurata*), the other species in its catalogue – in conditions closely approximating their natural habitats. Deep and spacious enclosures mean that the fish only account for 1 percent of the volume of their pens, reducing their density to one third that of conventional farms. Moreover, the clear waters of the Mediterranean flowing through the Gulf of Argostoli at Kefalonia Fisheries' production site are free of any traces of industrial, agricultural or urban activities and virtually unexposed to pollutants. No additives, antibiotics or chemicals

©Kefalonia Fisheries

> "Kefalonia Fisheries was founded with the aim to preserve the traditional way of life of the island through an innovative and sustainable professional practice: aquaculture."

©GFCM/Daniel Gillet

are used in Kefalonia's production cycle, allowing the fish to grow and mature at their own pace, as nature intended.

Reliably obtaining a top-end seafood product from its natural environment, however, is not as easy as it may sound. Kefalonia counts on a qualified team of scientists and trained personnel to ensure high survival and fertility rates of the fish from their conception and birth at the onshore hatchery through their transition to life in ocean waters. In the early days of the business, when no other Greek fish farms were producing seabass or seabream, it was essential to adapt lessons learned from raising other fish to these species.

A challenge that has accompanied Kefalonia Fisheries over the last four decades is the social acceptability of aquaculture. As with any growing industry, aquaculture faces resistance from those unfamiliar with the novel facilities,

technologies and products it brings. Local residents may worry about the effects of aquaculture on other coastal economic activities, such as tourism, and seafood consumers can be prejudiced to viewing farmed products as inferior to wild catch.

Responsible and sustainable producers like Kefalonia Fisheries have achieved great progress in relieving the concerns of skeptics in their own regions and around the world. Increasingly recognized as a boon for local economies and ecosystems by offering employment and development opportunities, protecting scarce natural resources and contributing to food security, the aquaculture sector finds an eager market worldwide. Kefalonia has developed strong relationships with a network of clients across fifteen countries and four continents.

Giving back to the community and helping to disseminate sustainable aquaculture practices are other important themes in Kefalonia Fisheries' story. The company has donated educational and laboratory equipment to local schools and taught valuable environmental lessons to young students through presentations and the organization of beach clean-up events. The name and ground-breaking success of Lara Barazi-Geroulanos is known and admired far beyond her small island. She is on the board of directors of the Hellenic Aquaculture Producers Organization (HAPO) and in 2020, she became the first woman to be elected president of the Federation of European Aquaculture Producers (FEAP).

©Kefalonia Fisheries

> "Fish consumption is popular in Algeria, though the species and style of cooking differ from one region to another. Couscous Cherchell is a popular dish in Cherchell, a city in the east of Algeria. It is made of couscous topped with a mixture of tomato, carrot, potato, garlic, cumin and pepper, and finally a piece of seared seabass."
>
> **Sarah Fodil**, Culinary Student at Institut Paul Bocuse.

Traditions
and recipes from the region

Seabass is considered one of the culinary pillars of the Mediterranean Sea and its recipes have been found in cuisines throughout the region for thousands of years. In the fourth century BCE, the gourmand Sicilian poet Archestratus of Gela, known as "the Daedalus of tasty dishes," praised seabass baked inside fig leaves and seasoned with olive oil and vinegar. Today, European seabass has become a pricy fish found in higher-end coastal restaurants, usually grilled whole or filleted and pan-fried.

Croatia / *Riba sa Žara*

In Croatia, seabass is grilled whole on an open fire, giving it a distinctive smoky aroma. First, the fish is gutted and scaled and the skin is cut with shallow incisions to prevent tearing during the cooking process. It is then seasoned with fine salt and drizzled with olive oil. Before putting the fish on the heat, the grill must be heated on an open fire until it is glowing hot to prevent sticking. The grilled seabass is usually accompanied by potatoes and spinach cooked in olive oil and by fresh lemon slices, which are squeezed on top at the last minute.

France / *Loup en croûte*

This traditional French dish made famous by the chef Paul Bocuse consists of a stuffed seabass cooked in a puff pastry. The fish is scaled and gutted and the fishbones are removed without detaching the head and the tail from the fillets. A fish mousseline seasoned with pistachios and tarragon is stuffed in between the two fillets. The fish is then covered with two pastry sheets carefully cut in the shape of a fish, highlighting the scales and the fins. After baking in an oven, the fish is cut in slices and served with a hollandaise sauce seasoned with tomato paste and tarragon and known as *sauce choron*.

Tunisia / *Hout melah* and *chermoula*

In the region of Sfax, *chermoula* is traditionally prepared for Eid al-Fitr, celebrating the end of Ramadan. The sauce is made of onion confit, hydrated raisins, olive oil, cumin, cloves and cinnamon and flavoured with mastic, a resin obtained from the mastic tree. The sauce ingredients are minced and served accompanied by salted and steamed white fish, which often includes seabass, called *hout melah*. The traditional recipe of *chermoula* from Sfax should not be confused with the Moroccan sauce of the same name, which is made of cilantro, parsley, garlic and chili and olive oil and used as a marinade for fresh fish, poultry and vegetables.

Chef's tips

How to prepare seabass?
As seabass is an easy fish to fillet, it is preferable to purchase a whole fish. To guarantee its freshness, look at the eyes: they should be translucent and shiny. Before cooking the fillets, sprinkle fine salt on the fleshy side and dry off any moisture with a clean towel. This technique will both season the fillets as well as prevent them from releasing water while pan-frying.

How to cook seabass?
Seabass flesh is delicate and can easily become dry when overcooked. Therefore, it is best to pan-fry, grill or roast the whole fish or fillets instead of poaching or using them in the preparation of a stew. Shallow incisions in the skin will prevent it from curling or cracking during the cooking process.

How to season seabass?
The flavours of seabass are delicate and easily over-seasoned. Therefore, focus on using fresh herbs and citrus zest instead of stronger spices. To intensify the taste of seabass in a dish, use the head and the fishbones as a base for a broth that can be reduced and emulsified with olive oil.

Here are the best ingredients to pair with seabass for unique Mediterranean and Black Sea-inspired dishes

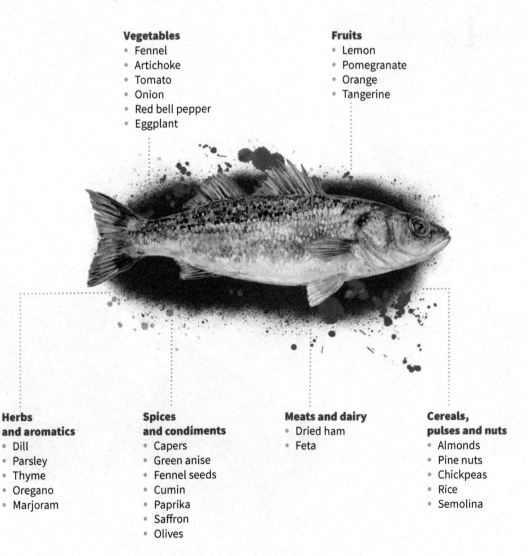

Vegetables
- Fennel
- Artichoke
- Tomato
- Onion
- Red bell pepper
- Eggplant

Fruits
- Lemon
- Pomegranate
- Orange
- Tangerine

Herbs and aromatics
- Dill
- Parsley
- Thyme
- Oregano
- Marjoram

Spices and condiments
- Capers
- Green anise
- Fennel seeds
- Cumin
- Paprika
- Saffron
- Olives

Meats and dairy
- Dried ham
- Feta

Cereals, pulses and nuts
- Almonds
- Pine nuts
- Chickpeas
- Rice
- Semolina

Caponata-stuffed seabass baked in a bread crust with olive and tomato condiment

Type
main course
Yield
4 servings
Preparation time
30 minutes
Cooking time
50 minutes

In this dish, seabass is stuffed with a traditional Sicilian caponata made with eggplants, tomatoes, olives and capers and baked in a bread crust to keep it tender and moist. For freshness and acidity, the dish is served with a tomato and olive condiment inspired by the sauce vierge from the south of France.

Utensils needed
- Cutting board
- Non-stick pan
- Fishbone tweezers
- Scissors
- Plastic wrap
- Mixing bowl
- Towel
- Rolling pin
- Baking tray

Ingredients

Caponata
- 2 eggplants
- 2 celery stalks
- 6 dried tomatoes
- 1 white onion
- 1 tbsp extra virgin olive oil
- 1 jar of capers
- 1 jar of green pitted olives
- 1 bunch of basil
- Salt and pepper

Stuffed seabass
- 2 pieces of seabass

Bread crust
- 350 g flour, type 55
- 10 g fine salt
- 15 g fresh baker's yeast
- 200 ml water
- 10 g dried thyme

Olive and tomato condiment
- 2 tomatoes
- 2 spring onions
- 2 garlic cloves
- 1 bunch of basil
- 2 tbsp taggiasca olives
- Juice of 1 lemon
- 2 tbsp extra virgin olive oil
- Salt and pepper

Preparation

Caponata
Finely dice the eggplants, celery, tomatoes and onion. In a pan, fry the onion in olive oil. Add the eggplants, celery and tomatoes and cook for 25 minutes on low heat. When the eggplants are tender, add the capers, green olives and basil leaves. Season to taste with salt and pepper and let cool.

Stuffed seabass
Remove the scales, fins and gills of the seabass. Remove the spine of the fish, keeping the fillets attached to the head and the tail. Remove the bones and rinse the fish under cold water. Stuff the seabass with the caponata, tightly encase in plastic wrap and refrigerate.

Bread crust
In a bowl, mix the flour and salt until combined. Add the yeast, water and thyme and knead for 5 minutes. Cover with a towel and let rest for 30 minutes at room temperature. Divide the dough into two equal portions and roll each to a thickness of 1 cm in the rough shape of the fish. Unwrap the seabass and place it on the first piece of dough. Cover with the second piece of dough and press the edges with a fork to seal. Bake in an oven preheated to 180 °C for 25 minutes, until golden-brown.

Olive and tomato condiment
While the fish is baking, peel and chop the onions and garlic, cut the tomatoes into small cubes, and rinse and finely chop the basil. In a bowl, mix the tomatoes, onions, garlic, basil, olives, lemon juice and extra virgin olive oil. Season with salt and pepper.

To serve
Slice the breaded seabass and serve with the olive and tomato condiment.

Nutrition facts

	Per 100 g	Per recipe
Energy	418 kJ/100 kcal	9 498 kJ/2 270 kcal
Protein	4.2 g	94.9 g
Carbohydrate	13.5 g	306.0 g
Fibre	2.4 g	53.9 g
Sugar	1.7 g	37.6 g
Fat	2.6 g	59.4 g
Saturated fat	0.4 g	9.5 g
Sodium	324 mg	7 315 mg

Rainbow trout

Oncorhynchus mykiss

Rainbow trout in the Mediterranean and Black Sea region

Scientific name:
Oncorhynchus mykiss

Family:
Salmonidae

Approximate average production volume (2016–2020):
259 000 tonnes

The top three producers are **Türkiye**, the **Russian Federation** and **France**.

96% of production occurs in **freshwater** and **4%** in **marine water**.

It is the most popular trout for culinary applications thanks to its **delicate flavour** profile.

Source: FAO. 2023. Global aquaculture production quantity (1950–2020). In: *Fisheries and Aquaculture Division*. Rome. Cited 26 July 2022. fao.org/fishery/statistics-query/en/aquaculture/aquaculture_quantity

Based on data from the GFCM Information System for the Promotion of Aquaculture in the Mediterranean (SIPAM)

Oncorhynchus mykiss

Rainbow trout

The most commonly farmed trout worldwide is the rainbow trout (*Oncorhynchus mykiss*). This has been a favoured species of aquaculture producers around the globe since the nineteenth century thanks to its reputation as a species that is hearty, fast-growing, easy to spawn and tolerant to a wide range of environments (FAO, 2022i; Woynarovich *et al.*, 2011). It presents a beautiful olive-green colouration with a pink band on either side and small black spots. In order for this fish to properly grow, high-quality waters are necessary and for this reason, it can be considered an indicator of the healthiness of a water basin (Pander *et al.*, 2009).

Culinary and nutritional value

Trout is commonly eaten near rivers in the Mediterranean and Black Sea region. Rainbow trout, boasting a delicate flavour profile, is the most popular variety of trout for culinary applications. In addition to being an excellent source of omega-3 fatty acids and a significant quantity of minerals, trout is a semi-fatty fish. Prepared whole, it can be stuffed and grilled or baked in a salt or bread crust. Trout fillets can be cured and smoked and eaten in thin slices on toast.

Farming in the region

Indigenous to the Pacific drainages of North America, rainbow trout has since been introduced to waters on all continents besides Antarctica for aquaculture purposes. Production of the species significantly expanded following the introduction of pelleted foods in the 1950s (FAO, 2022i). Today, production of rainbow trout in the Mediterranean and Black Sea region is highest in Türkiye, the Russian Federation, France and Italy, which together produced approximately 265 000 tonnes in 2020 (FAO, 2023). The species is generally produced on its own in an intensive system using long concrete tanks called raceways or ponds; less commonly, cages and recirculating systems are used. Rainbow trout do not spawn naturally in aquaculture systems, and therefore the production cycle begins by artificially spawning eggs from high-quality brood fish (FAO, 2022). The eggs, characterized by a yellow-orange colouring due to the antioxidants and carotenoids in the yolk (Craik, 1985), are then incubated in hatching troughs for anywhere from 21 to 100 days, depending on the temperature. When the juvenile trout, called fry, reach 8–10 cm, they are transferred to grow-out facilities to grow to commercial size, a process that typically takes 9 months. At that time, producers are able to begin the harvest, working to keep stress on the fish to a minimum.

Kuzuoğlu Aquaculture: a son of the Çağlayan valley returns with a visionary spirit

The cool, highland waters of Türkiye's Çağlayan Valley offer the ideal conditions for raising trout, a fact that Hasan Kuzuoğlu recognized while he was working in the American tourism sector, importing salmon from Chile on behalf of his employer. Intrigued and encouraged by the global demand for this fish, he returned to Türkiye and the waters of the Çağlayan Valley, with a vision and strong business acumen in tow, to revive his family's idle farm and shift its focus to salmon and trout production.

In 2011, he secured a substantial investment for the farm and established its first large, integrated facility. That year, 3 tonnes of fish were produced; the next season saw a substantial increase to 25 tonnes, while subsequent years recorded 75 tonnes, 150 tonnes and 500 tonnes, with production reaching 15 000 tonnes by 2022. Consumers in over ten countries are continually impressed by the unique taste and high quality of Kuzuoğlu's products. The primary subject of this production is rainbow trout, marketed as Turkish salmon, which accounted for 13 500 tonnes of the 2022 production volume of the farm.

Kuzuoğlu proudly processes its products at its own state-of-the-art 5 000 m² facility, which is staffed by numerous teams of professionals following the operations each step of the way. The advanced facility is capable of processing 120 tonnes of gutted, filleted and smoked fish each day.

At each stage in the process, the enterprise takes into consideration the environmental impacts of its work and actively chooses to operate following the principle of sustainability. In addition to conventional fish farming methods, Kuzuoğlu has adopted, and continues to adapt, recirculating aquaculture systems. These systems filter and recirculate the water used for farming, reducing Kuzuoğlu's need to source clean water while still maintaining a suitable environment for the fish. Meanwhile, at the processing plant, Kuzuoğlu takes a sustainable approach to its waste management by making fishmeal from leftover materials and treating any liquid waste that cannot be recycled.

To continue on the path of substantial growth, efforts are currently underway to enlarge the existing cages and increase the volume of equipment available. The enterprise also acknowledges the importance of improving the social acceptability of aquaculture, which it recognizes is a key factor in increasing domestic demand for the products and enticing additional investments.

©Kuzuoğlu Aquaculture

"Our motivation is the trust and satisfaction of our customers."

©Kuzuoğlu Aquaculture

"Seafood and fish, like trout, are commonly eaten in Lebanon. A popular dish is *sayadieh*: oven-roasted trout with lemon served atop rice cooked in a variety of spices including turmeric, coriander and cinnamon and garnished with caramelized onions. Salad and a tahini paste called *tarator* can be added on the side."

Chaden Ziadeh, Culinary Student at Institut Paul Bocuse

Aquafarm: circling self-sufficiency

Moving from the Çağlayan Valley to the Lebanese coast, we find Aquafarm, an aquaculture enterprise endeavouring to enhance the sustainability of aquaculture and promote aquaculture initiatives in the region.

Originally founded by Massaad Ejbeh in 2001, Aquafarm began as a tilapia and catfish farm. Two years later, it expanded to include shrimp farming in its catalogue. The success of this shrimp farming venture pushed Ejbeh to make it the entire focus of Aquafarm and for the next 15 years, he and his team produced a healthy, affordable protein for local markets. This progress was abruptly interrupted in 2019 when economic crisis hit Lebanon hard, and forced Ejbeh and his team to rethink their business.

The first steps in their plan of action were to supplement their shrimp farming practices with trout, mullet and salmon and to re-establish tilapia production. The next phase was to increase the self-sufficiency of the enterprise in order to better insulate themselves against rising costs and the dollar shortage. This meant establishing an in-house feed mill to secure a steady supply of low-cost feed and eliminate the need for imports and transitioning its water pumps from oil to renewable energy sources.

Through each of these initiatives runs a common thread of sustainability. Not only has the switch to renewable sources meant a greater level of self-sufficiency, but also a reduction in emissions. Meanwhile, at its feed mill, Aquafarm uses only natural ingredients, many of which are sourced from the farm itself. Ejbeh and his team are working to improve the enterprise's waste management practices and are transforming the leftovers from their salmon processing into salmon oil and salmon meal, which are then used to make pet food.

Given the ongoing financial crisis, political instability and rising costs in Lebanon, Ejbeh is working towards achieving a completely self-sufficient farm. He envisions a circular environment in which all projects and activities fulfil each other, all under the umbrella of sustainability. With these actions, he aims to secure the future of Aquafarm, allowing it to continue to operate as a provider of employment and healthy and affordable food and as an example of sustainability in the region.

"The dream is to be completely self-sufficient and create a circular environment through our activities on the farm."

Traditions
and recipes from the region

The name trout derives from the Greek *troktis*, meaning voracious fish. The ancient Greeks and Romans already developed the technique of smoking trout fillets to aid in the preservation of the fish as well as to develop its flavours. Today, smoked trout is promoted as an alternative to smoked salmon in order to favour fish caught in rivers feeding into the Mediterranean and the Black Sea over those imported from northern Europe. Freshly caught, trout is consumed pan-fried, baked with sauces or featured in soups.

©GFCM/Claudia Amico

Italy / *Trota alla valdostana*

This traditional recipe from the Valle d'Aosta region in the Alps is prepared with trout freshly caught from a local river. It is gutted and rinsed and, in the meantime, carrots, celery and onions are diced into small pieces and pan-fried with sage and rosemary. The trout is added to the pan and coloured on both sides before being deglazed with white wine vinegar and cooked in a fish broth seasoned with rehydrated raisins and lemon zest. When the flesh is cooked through, the broth is reduced and emulsified with butter to be served as a sauce along with the trout.

©GFCM/Dominique Bourdenet

Morocco / **Trout *briouates***

Also known as *brioute* in Türkiye, *boubek* in Lebanon and *brique* in Tunisia, this filled brick pastry is usually prepared with ground meat and eaten at family celebrations. In the Moroccan Atlas Mountains, a fish version of *briouate* is prepared with rainbow trout farmed in the nearby rivers. The trout is smoked and mixed with fresh cheese and wrapped in a sheet of brick pastry, baked until crisp and golden and served with raw vegetables and lemon slices.

©GFCM/Dominique Bourdenet

Slovenia / *Postrv v ajdovi ali koruzni moki*

Traditionally prepared with native marbled trout from the Soška River, a species currently endangered, the recipe has been adapted nowadays to use rainbow trout. The fish is gutted and scaled, then breaded with buckwheat and cornmeal. It is pan-fried at a moderate temperature until the flesh is cooked through and the cornmeal crust turns golden-brown. The trout is served with lemon slices, potatoes and spinach.

Chef's tips

How to prepare trout?
Trout can be used in a multitude of culinary preparations, from raw tartare to grilled whole on an open fire. One method of preparation is to fill it with vegetables or a mousseline. To prevent the filling from spilling out during the cooking process, the stuffed trout can be rolled in plastic wrap and poached for 30 seconds in simmering water, then unwrapped and pan-fried or roasted.

How to cook trout?
Prepared whole or in fillets, trout's fatty flesh withstands longer cooking periods than the flesh of white fishes. This advantage offers the possibility to roast, grill, poach, pan-fry, deep-fry or even bake trout in a salt, bread or pastry crust. Trout fillets can also be cured with salt, sugar and spices and served raw in thin slices on toast.

How to season trout?
Trout's rich aromatic profile pairs ideally with aniseed herbs and vegetables such as fennel, green anise or dill. It can also be served as a soup with perfumed spices, nuts and dried fruits. When poached, the skin of the trout can be sauteed with olive oil to intensify the fish's flavour and then crushed with spices to add as a topping to a dish.

Here are the best ingredients to pair with trout for unique Mediterranean and Black Sea-inspired dishes

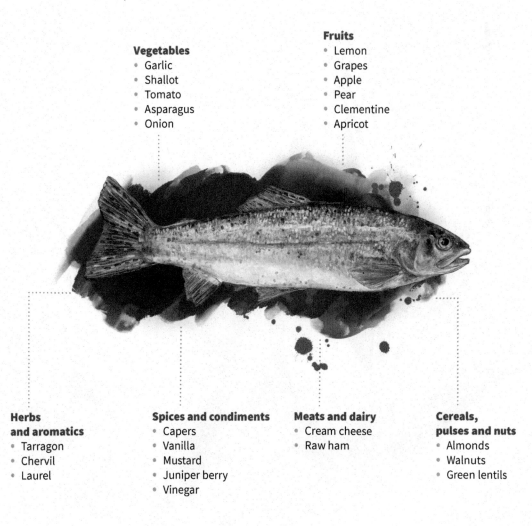

Vegetables
- Garlic
- Shallot
- Tomato
- Asparagus
- Onion

Fruits
- Lemon
- Grapes
- Apple
- Pear
- Clementine
- Apricot

Herbs and aromatics
- Tarragon
- Chervil
- Laurel

Spices and condiments
- Capers
- Vanilla
- Mustard
- Juniper berry
- Vinegar

Meats and dairy
- Cream cheese
- Raw ham

Cereals, pulses and nuts
- Almonds
- Walnuts
- Green lentils

Baked trout with sage served with a zucchini and walnut condiment

Type
starter
Yield
4 servings
Preparation time
15 minutes
Cooking time
40 minutes

This recipe pairs the delicate flesh of trout with the herbaceous aroma of sage. The whole fish is baked atop a layer of baby potatoes that confit in trout fat. The dish is served with a fresh condiment made with finely diced zucchini and walnuts.

Ingredients

Confit baby potatoes
- 500 g baby potatoes
- 2 garlic cloves
- 2 sage stems
- 4 tbsp extra virgin olive oil
- Salt and pepper

Baked trout with sage
- 1 rainbow trout
- 2 garlic cloves
- 2 sage stems
- Salt and pepper
- 2 tbsp extra virgin olive oil

Zucchini and walnut condiment
- 2 shallots
- ½ bunch of basil
- 2 zucchini
- 1 handful of walnuts
- 3 tbsp extra virgin olive oil
- Salt and pepper

Utensils needed
- Cutting board
- Baking dish
- Fishbone tweezers
- Mixing bowl

Preparation

Confit potatoes
Wash the potatoes and cut the bigger ones in two. Place into a large baking dish with the garlic, sage and olive oil. Season with salt and pepper.

Baked trout with sage
Rinse the fish under water, remove the scales, gills and fins and rinse again. Cut shallow incisions into the skin to prevent it from cracking while cooking. Stuff the trout with garlic and sage leaves. Season with salt and pepper and sprinkle with olive oil. Place the trout on top of the potatoes and bake in an oven preheated to 180 °C for 40 minutes.

Zucchini and walnut condiment
While the trout is baking, peel and chop the shallots, finely slice the basil leaves, finely dice the zucchini and chop the walnuts into small pieces. In a bowl, mix the shallots with the basil, zucchini, walnuts and olive oil. Season with salt and pepper.

To serve
Place the fish on top of the confit potatoes and top with the zucchini and walnut condiment.

Nutrition facts

	Per 100 g	Per recipe
Energy	552 kJ/132 kcal	13 359 kJ/3 193 kcal
Protein	7.1 g	172.0 g
Carbohydrate	3.8 g	92.1 g
Fibre	1.6 g	39.2 g
Sugar	2.7 g	64.9 g
Fat	9.4 g	226.0 g
Saturated fat	1.3 g	32.3 g
Sodium	58.6 mg	1 417 mg

Pacific oyster

Magallana gigas

Pacific oyster in the Mediterranean and Black Sea region

Scientific name:
Magallana gigas

Family:
Ostreidae

Approximate average
production volume
(2016–2020):
84 800 tonnes

The top three producers are
France, **Spain** and **Morocco**.

99% of production
occurs in **marine water** and
1% in **brackish water**.

A prized mollusc with
a long history in the region,
perfect for festive occasions.

Source: FAO. 2023. Global aquaculture production quantity (1950–2020).
In: *Fisheries and Aquaculture Division*. Rome. Cited 26 July 2022.
fao.org/fishery/statistics-query/en/aquaculture/aquaculture_quantity

Based on data from the GFCM Information System for
the Promotion of Aquaculture in the Mediterranean (SIPAM).

Magallana gigas

Pacific oyster

The Pacific oyster (*Magallana gigas*), the most famous member of the Ostreidae family and the most farmed bivalve in the world, is known for its rapid growth and resilience against varying environmental conditions. These hearty characteristics have allowed producers to establish oyster farms in areas where they did not previously exist, leading to a flourishing aquaculture sector, in particular in France (FAO, 2022j; Turolla, 2020). Farming of the Pacific oyster has also allowed for the support of severely depleted wild stocks (FAO, 2022j).

Culinary and nutritional value

Oysters are a prized mollusc commonly consumed in Europe on festive occasions such as Christmas and New Year's Day. Their flesh has a delicate taste profile with nutty and iodine flavours and is rich in protein, amino acids, vitamins and minerals. Usually consumed raw with lemon juice or vinegar, they can also be served warm, such as poached, gratinated or fried. Oysters have become a source of inspiration in higher-end restaurants, which are developing innovative ingredient pairings and reimagining methods of cooking.

Farming in the region

Though the Pacific oyster originated in the waters near Japan, the Republic of Korea, the Democratic People's Republic of Korea, northern China and the Russian Federation, it can now be found along the northeast Atlantic coast from Scandinavia to North Africa as well as in the Mediterranean and the Black Sea (Alvarez *et al.*, 1989). Production in this region dates back to the Roman Empire (Günther, 1897), when oysters were transported from natural beds to saltwater lakes along the southern Italian coast to create artificial reefs for easy harvest (Bardot-Cambot and Forest, 2013). It was not until the second half of the

twentieth century that the modern expansion of this practice started on a global scale (Botta *et al.*, 2020). Today, oyster aquaculture in the region is concentrated in France, which accounts for more than 90 percent of production (EUMOFA, 2022). However, oyster production initiatives are quickly spreading throughout the basin, making it possible to find oysters in restaurants and markets in numerous countries in the region.

Production methods vary depending on the environment, though they all begin with the establishment of a seed, also known as spat, supply. Around the world, oyster aquaculture is mainly based on natural spat collection;

however, producers also have the option to source spat from oyster hatcheries, with many having been opened in recent years (FAO, 2022j). Once the spat supply has been established, the ongrowing process begins. This process occurs mainly at sea and can involve a variety of culture methods including on-bottom, off-bottom, suspended and floating systems. While oysters can grow to a size of about 100 mm in shell length, Pacific oysters are typically marketed from 60 mm in shell length upwards and it is at this size that harvest of the species begins.

Ostras de Valencia: captivating chefs with unique oysters

What do each of the Michelin-starred restaurants in Valencia, Spain have in common? The oysters they serve to their guests have all passed through the hands of the farmers at Ostras de Valencia. Grown at depth, handled individually and differing in variety from the other oysters in the region, Ostras de Valencia oysters are also known as the "pearls of Valencia" and are famous among local chefs for their size, shape and flavour.

This fame is particularly notable given the enterprise's establishment less than two decades ago, in 2007. The idea started with César Gómez, an experienced fisher who set out to draw on his years of expertise and produce oysters in the Ebro Delta. For three years, he successfully grew his operations in the area until 2010, when he decided to shift his focus to cultivating oysters at the port of his home city, Valencia. His plan centred around using the same trays for oysters as were already used for mussel production – something that had never been done in the area before. Gómez was optimistic

that the high-quality waters and excellent climatic conditions that lend themselves to producing the popular Valencian *clóchina* – a local variety of Mediterranean mussel (*Mytilus galloprovincialis*) – would help to create an exceptional oyster. He was correct, and though it took some effort, considering the uniqueness of oyster production in the port of Valencia, to convince local chefs of the quality of the Ostras de Valencia product, one taste and they jumped at the opportunity to serve these oysters in the area's top restaurants.

Traditionally, European flat oyster (*Ostrea edulis*) has been the species of choice for producers in Spain, but the team at Ostras de Valencia wanted to present a quality product that would stand out from the rest. So they turned to *Magallana gigas*, an oyster species that is native to the Pacific but acclimatized long ago to European waters. Compared to the flat oyster, the Pacific oyster

is plumper with a more sharply pointed shell and a less briny flavour. It has become famous worldwide thanks to its use in French cuisine, with chefs drawn to its delicate flavour and vegetable-like finish.

Ostras de Valencia's strength lies not only in the creative use of a new species in a new location, but also in the artisanal approach it takes to production. The seven-person team works manually, carefully handling each oyster one by one from seed to final harvest. This process begins with the arrival of high-quality oyster seeds meticulously selected from France when they are the size of mere grains of rice. For the next four months, the seeds are cultivated in lanterns known as *cubanitos* until they reach an adequate size. Then comes the moment to sow the seeds to the ropes suspended under the tables located in the port of Valencia. This process has largely remained the same since 2010 when operations began, but the tray has since been adapted for more comfortable work, including a larger work surface and additional supports to walk between the boards.

> "The aim since the beginning has been to offer a high-quality product that stands out from the rest."

©Ostras de Valencia

The oysters spend roughly one year in the water. During this time, Ostras de Valencia does not employ any chemical products or feed, instead allowing the oysters to feed on the phytoplankton in the water. This practice, known as affinage, allows Ostras de Valencia to achieve highest-quality shells, meat and colour. It also allows the enterprise to minimize its impacts on the environment surrounding the tables and to improve the sustainability of their work.

Since César Gómez first set out on his mission to farm oysters in the port of Valencia in 2010, the production site has grown to cover 900 m² and production has grown to a volume of 3 tonnes annually. Ostras de Valencia has no plans to stop here. They aim to continue growing their operation by adding new facilities, new boats and more staff in order to expand production, enhance their supply nationally and continue providing restaurants and municipal markets with an exceptional-quality oyster.

"In the south of France, we love oysters served with red wine vinegar, rye bread and salted butter. This shellfish can also be used to prepare a popular holiday dish: a smooth sauce made of egg yolks, champagne, butter and cream is poured atop oysters in their shells, which are then baked in the oven for five minutes."

Lisa Barboteu, Culinary Student at Institut Paul Bocuse

Traditions
and recipes from the region

Oysters were appreciated and consumed in abundance by the ancient Romans, the Celts and the Greeks. Guillaume Tirel "Taillevent", chef at the medieval French court of the early Valois kings, mentioned four ways of preparing oysters, including by battering and frying, in his culinary book *Livre fort excellent de cuysine très utille et proffitable, contenant en soy la manière d'abiller toutes viandes, avec la manière de servir ès banquetz et festins, le tout veu et corrigé oultre la première impression par le grant escuyer de cuysine du Roy*. Traditionally eaten for winter festivities in European countries, oysters are paired with white wine or champagne to enhance their delicate flavour.

France / *Huîtres gratinées à la persillade*

It is a Christmas and New Year's Eve tradition in France to begin dinner with raw oysters served with a glass of white wine along with toasts or rye bread. However, in some families, oysters are served warm with a parsley and garlic breading. The oysters are first opened and rinsed under clear water. Stale bread is dried in the oven at a low temperature and then blended with parsley leaves and fresh garlic cloves. The oysters are covered with the breading while still in their shells, each topped with a piece of butter and baked until golden-brown.

Italy / *Zuppa di ostriche*

In Italy, oysters feature in a traditional Christmas soup recipe. Pancetta, onions, celery and leeks are pan-fried before tomatoes and peppers are blanched and added to the broth along with fresh herbs and potatoes. The soup is simmered at low heat until the potatoes are cooked through. While cooking, the oysters are opened and their juice is filtered to remove any pieces of shell. The soup is taken off the heat and just prior to serving, the oysters are added to poach them gently.

Spain / *Ostras al cava*

This festive Spanish recipe combines oysters with a sparkling wine called cava. Onions and shallots are chopped and sauteed in a pan and deglazed with the sparkling wine. The sauce is then reduced with cream and a pinch of freshly ground black pepper. The oysters are opened, rinsed under clear water and topped with the sauce before being gratinated under a grill. This recipe can also be prepared with diced tomato added to the oysters before they are topped with the sauce.

Chef's tips

How to prepare oysters?

Opening oysters can be risky without the right technique. Use a short knife specifically designed for shellfish and protect your hand by holding the oyster with a thick towel. If the oysters are still sandy, remove their liquid and rinse them with cold water. An alternative to opening the oysters with a knife is to steam them for 20 seconds – the shell will open with slight pressure.

How to cook oysters?

Usually prepared at the last minute and eaten raw, oysters are also an excellent seafood when served warm. Similar to mussels, oyster flesh can be pan-fried, breaded and fried or served in a stew. Fast cooking processes are preferred as the texture can become chewy when cooked for too long. When preparing oysters in a broth or a soup, make sure to never bring the liquid to a boil once the oysters have been added.

How to season oysters?

The flavours of raw and cooked oysters are different, and pairings should be adapted accordingly. When eaten raw, the flavour profile is enhanced by the use of acidic ingredients like lemon, green apple or vinegar. When cooked, oysters can be paired with rounder aromatics such as tomato sauce, sauteed leeks or smoked paprika.

Here are the best ingredients to pair with Pacific oyster for unique Mediterranean and Black Sea-inspired dishes

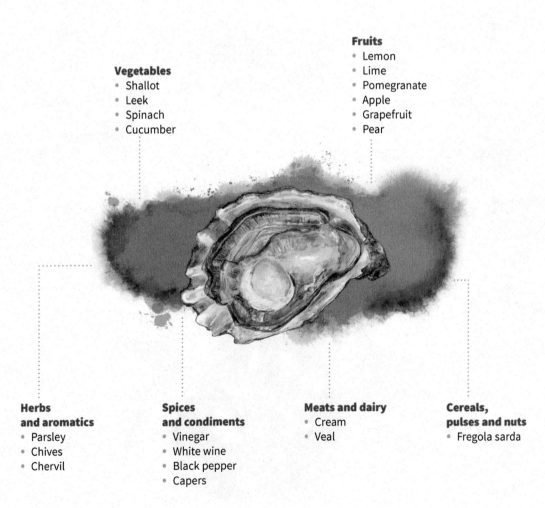

Vegetables
- Shallot
- Leek
- Spinach
- Cucumber

Fruits
- Lemon
- Lime
- Pomegranate
- Apple
- Grapefruit
- Pear

Herbs and aromatics
- Parsley
- Chives
- Chervil

Spices and condiments
- Vinegar
- White wine
- Black pepper
- Capers

Meats and dairy
- Cream
- Veal

Cereals, pulses and nuts
- Fregola sarda

Oyster, peach and tomato tartare

Type
starter
Yield
4 servings
Preparation time
20 minutes
Cooking time
20 minutes

Serving oysters as a tartare instead of raw is a great way to present this ingredient to a larger number of people. The iodine flavour of the oyster pairs well with the sweetness of the peaches and the freshness brought by the tomatoes. Feta cheese and roasted almonds finish the dish.

Ingredients

Oyster, peach and tomato salad
- 8 pacific oysters
- 2 red tomatoes
- 1 yellow tomato
- 1 yellow peach
- 2 mint leaves
- Zest and juice of 1 organic lemon
- 2 tbsp extra virgin olive oil
- Salt and pepper

Topping
- 1 handful of almonds
- 1 tbsp feta cheese

Utensils needed
- Oyster knife
- Cutting board
- Bowl
- Mixing bowl

Preparation
· · · · · · · · · · · · · · ·

Oyster, peach and tomato salad

Open the oysters with an oyster knife and rinse under cold water to remove any pieces of shell. Cut the oyster flesh into small cubes, place them in a bowl and refrigerate. Clean and dry the oyster shells.

Finely dice the peach and the tomatoes and finely chop the mint. In a bowl, mix the oyster tartar with the tomatoes, peaches, mint, lemon zest, lemon juice and olive oil. Season with salt and pepper and refrigerate.

Topping

Roast the almonds in a preheated oven at 170 °C for 10 minutes and chop them in small pieces. Crumble the feta into small pieces.

To serve

Divide the salad among the cleaned oyster shells and top with the chopped almonds and crumbled feta cheese.

Nutrition facts

	Per 100 g	Per recipe
Energy	302 kJ/72 kcal	4 644 kJ/1 110 kcal
Protein	5.2 g	79.4 g
Carbohydrate	3.9 g	59.3 g
Fibre	0.7 g	10.9 g
Sugar	1.6 g	25.1 g
Fat	3.8 g	57.8 g
Saturated fat	0.7 g	10.9 g
Sodium	323 mg	4 971 mg

©GFCM/Nicolas Villi...

Beluga sturgeon

Huso huso

Beluga sturgeon in the Mediterranean and Black Sea region

Scientific name:
Huso huso

Family:
Acipenseridae

Approximate average
production volume
(2016–2020):
14 tonnes

The top producer is **Bulgaria**.

98% of production occurs
in **freshwater** and
2% in **brackish water**.

While it is famous for its
roe, **its fillets are healthy,
versatile** and ideal for weekly
family meals.

Source: FAO. 2023. Global aquaculture production quantity (1950–2020).
In: *Fisheries and Aquaculture Division.* Rome. Cited 26 July 2022.
fao.org/fishery/statistics-query/en/aquaculture/aquaculture_quantity

Based on data from the GFCM Information System for
the Promotion of Aquaculture in the Mediterranean (SIPAM).

Huso huso

Beluga sturgeon

Known as "living fossils", sturgeons are the oldest freshwater fish in the world, with roots dating back 200 to 250 million years. These fish differ from the other species encountered in this guide in several key ways: their skeleton is made of cartilage instead of bones, they can live up to 100 years and they can grow over 6 m long (WWF, 2020; FAO, 2013). The name sturgeon covers a group of 27 species, seven of which are indigenous to the Black Sea, including the beluga sturgeon (*Huso huso*), and several others of which are indigenous to the Mediterranean basin (FAO, 2013). The wild populations are currently endangered. As such, aquaculture presently offers the only sustainable route for consumption of sturgeon meat or caviar, leading to its widespread culture.

Culinary and nutritional value

Sturgeon is a storied fish across the northern regions of the Mediterranean and the Black Sea, long sought after for its roe, which provides the basis for caviar. Its dense flesh is of a pinkish colour and can be used in a variety of dishes: grilled on an open fire, pan-fried or poached in the preparation of soups. Rich in proteins and in vitamin B₃, sturgeon fillets are ideal for weekly family meals. The fillets are also prepared cured and smoked to aid in their preservation and to develop their flavour characteristics.

Farming in the region

With many sturgeon species endangered due to overfishing and pollution, aquaculture provides one of the means to enhance the production of caviar and an opportunity to reintroduce sturgeons into areas where their populations have decreased dramatically. In the Mediterranean and Black Sea region, sturgeon production exceeded 7 000 tonnes in 2020 and encompassed an incredibly diversified group of

species, including beluga sturgeon (FAO, 2023). The main producer in the region is the Russian Federation, followed by Italy, together accounting for more than 80 percent of the total production of sturgeon (FAO, 2023). This fish is anadromous, meaning that it migrates from saltwater to freshwater to spawn and can therefore be raised under different farming regimens. For instance, the entire production cycle may occur in freshwater, or it can begin in freshwater before moving to higher salinities. One of the main constraints on sturgeon culture is late sexual development, which for beluga sturgeon occurs after

16–18 years on aquaculture farms (EUMOFA, 2021). Today, thanks to innovations such as ultrasound technology, it is possible to evaluate gonadal development and extract the roe from the females without sacrificing the animals. In the Black Sea basin, to which many of these species are indigenous, many efforts are being carried out to further develop the sturgeon aquaculture sector and to reintroduce these fish into the wild (Massa *et al.*, 2021).

Caviar Giavari:
an ancient fish calls for
patience and wit

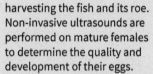

Renaissance Italians were great lovers of the delicate flesh and roe of sturgeons swimming in the Po river and its tributaries. The pope's private chef included both in original recipes and painters captured the impressive fish in fine detail. Legend has it that Leonardo da Vinci even filled a small gemstone-encrusted chest with sturgeon caviar to present as a wedding gift to Beatrice d'Este.

Long after the disappearance of these few indigenous species from Italy, the same cool waters of the Po valley that nurtured them are once again home to sturgeon. Recent decades have seen a handful of aquaculture farms along the Po begin raising sturgeon, supporting a young Italian caviar industry that is growing in size and global recognition. Leading the pack are the Giaveris – father Rodolfo and daughters Jenny, Giada and Joys – who breed ten species, including beluga sturgeon, in their 15-hectare system of ponds and tanks, located 20 km from Venice.

At the Giaveri farm, the geochemical properties of the waters emerging from underground springs are calibrated to ensure that the sturgeon experience the same conditions they would in their natural environments of the Black Sea, Caspian Sea and the rivers of the Russian Federation. Attention to the elements in which the sturgeons live, however, is only the first step in a process of individualized care for the fish, who can live over 20 years.

"The cycle of production begins with the egg and ends with the egg," the Giaveris write. As the sturgeons grow toward maturation, which they reach as early as age 7 and as late as age 15 depending on the species and the individual, it is essential to monitor each fish's feeding, behaviour and health. This supervision is performed by computer systems and scientific experts.

A combination of technology and human skill is also used in harvesting the fish and its roe. Non-invasive ultrasounds are performed on mature females to determine the quality and development of their eggs.

Fish with suitable roe satisfying Giaveri's high standards are treated to a process of gastronomic craftsmanship that combines the traditions of the Russian Federation with Italian mastery. The eggs are extracted carefully by hand and every impurity is washed away before the salting process begins. Giaveri proudly practices the time-honored salting method called *malassol* (meaning little salt in Russian), which launches a three-month aging process in vacuum-sealed tins that gives Giaveri's caviar its distinctive taste.

Caviar catapulted to the opulent status it enjoys today under the rule of tsars in what is now the Russian Federation. These sovereigns savoured the roe from abundant sturgeon populations in the Black Sea and the empire's many rivers. *Huso huso* roe, known as beluga caviar, rose to particular fame for its large grains and unique flavour. However, as caviar's reputation grew worldwide, particularly in the Near East and at European courts, it began to be exported by the tonne, resulting in overfishing of the Russian Federation's native sturgeon stocks and near extinction for many species. In the 1990s, coinciding with the dissolution of the Soviet Union, the Convention on International Trade in Endangered Species of Wild Flora and Fauna (CITES) placed heavy restrictions and even bans on wild sturgeon fishing (CITES, 2023).

By this time, Rodolfo Giaveri had already introduced sturgeon to the aquaculture farm he founded for breeding eels in the 1970s. His original purpose for adding sturgeon was for sport fishing and the production of a delicious fish that had, until recently, been native to the region. The global limits on wild sturgeon fishing, coupled

©GFCM/Daniel Gillet

"Sturgeon breeds will always hold a special place in the cuisine of the Russian Federation. This fish can be used to prepare a traditional soup called *ukha osetra*, for which a strong broth is brewed using the head of the fish along with carrots, onion and aromatic herbs. Diced potatoes and sturgeon pieces are then added and the soup is finished with garlic, parsley and dill."

Daniil Nikulin, Culinary Student at Institut Paul Bocuse

©Caviar Giaveri

©Caviar Giaveri

with continued demand for caviar, however, offered fish farmers an opportunity that Rodolfo was uniquely prepared to exploit.

Over the next twenty-five years of caviar production, the Giaveris have stayed true to the roots of this delicacy, organizing workshops for their employees on techniques and methodologies led by experts from the Islamic Republic of Iran and the Russian Federation, while also innovating and experimenting.

"Our activities are carried out with strict respect for nature and for maintaining the integrity of the environments where we raise the fish."

They have continued adding different sturgeon species to their pools, and now their farm boasts an incredibly diverse sturgeon stock.

According to gastronomic authorities, the term caviar is used to describe only roe from the *Acipenser* and *Huso* families, together comprising the sturgeon clade. However, there can be confusion among consumers about the origin of their caviar purchases.

Giaveri holds itself to the highest standards of traceability and quality, offering a clear and easily readable label on all its tins with a traditional blue hue and red ribbon. Beyond its efforts to ensure the comfort and health of its customers, Giaveri also strives to minimize its environmental impact, saving and recycling water and using only green energy from renewable sources to operate the farm. Because of breeding efforts by the Giaveris and others like them, scientists have even been able to reintroduce Adriatic sturgeons (*Acipenser naccarii*) into the Po watershed. They hope conditions will allow the majestic giant beluga sturgeon to one day return to the western edge of its ancient range as well.

Traditions and recipes from the region

During the Middle Ages, sturgeon was highly appreciated and served in various dishes at the tables of sovereigns. Before being prized for its roe, the sturgeon's fillets were cured to extend their preservation time. Caviar appeared around the ninth century CE when the practice of salting sturgeon eggs began. This refined food rapidly became widely popular and today, it is a delicacy served in higher-end restaurants.

Russian Federation / Caviar

Caviar is renowned as one of the most expensive delicacies in the world. It is prepared with sturgeon roe that is rinsed, soaked in a brine, carefully sorted and delicately packed in tin boxes. Caviar is traditionally served with vodka, but today it is most commonly paired with champagne and found in higher-end restaurants. The fillets of the farmed female sturgeon raised for caviar production are often sold at an affordable price and used in broths and soups.

Russian Federation / Smoked sturgeon

This culinary delight, made with the cured and cold-smoked meat of large female freshwater sturgeon, is served on toast with pickled vegetables. The fish is first gutted, filleted and rinsed with fresh water. It is then soaked in a brine with sugar and cold-smoked using applewood chips to preserve the flesh and develop its aromatic profile. In sturgeon-producing regions, smoked sturgeon is more affordable than smoked salmon and therefore provides an intriguing substitute for recipes.

Ukraine / *Ukha*

This traditional fish stock recipe is prepared with white fish, such as sturgeon, and served as a first course during festivities. Fish bones, tails and heads are delicately poached with onions, carrots, bay leaves, dill, tarragon, parsley, nutmeg and fennel seeds. The broth is strained and served with poached fish fillets, carrots and minced parsley. In some version of the dish, saffron pistils are added, offering a golden-orange colour to the broth.

Chef's tips

How to prepare sturgeon?
Sturgeon's flesh structure, similar to that of swordfish and yellowfin tuna, can easily dry out during the cooking process. To prevent this, fish fillets can be marinated in olive oil with spices and herbs for one hour. The fat brought to the fish by the olive oil will help to keep the flesh moist when it is grilled or pan-fried.

How to cook sturgeon?
Sturgeon is often found in fillets or as steaks with the backbone intact. To prevent the fish from losing its moisture during the cooking process, it can be baked in a salt crust. Wrap the fillets with vegetable leaves, such as Swiss chard, fig leaves or algae,

to prevent direct contact with the coarse salt. The crust can also be prepared out of bread or pastry and served with the fish. Sturgeon can also be smoked and used as a replacement for smoked salmon in the preparation of sandwiches or toasts.

How to season sturgeon?
The flesh of the sturgeon is aromatic and pairs well with fresh and acidic ingredients like balsamic vinegar, lemon, tarragon or sorrel. It can be also perfumed with light spices such as cumin, sumac or powdered ginger.

Here are the best ingredients to pair with sturgeon for unique Mediterranean and Black Sea-inspired dishes

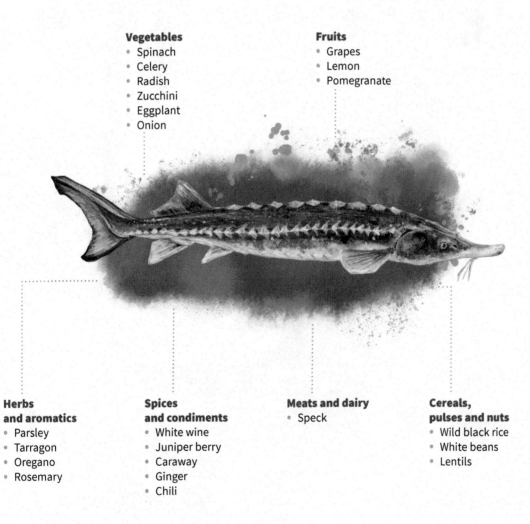

Vegetables
- Spinach
- Celery
- Radish
- Zucchini
- Eggplant
- Onion

Fruits
- Grapes
- Lemon
- Pomegranate

Herbs and aromatics
- Parsley
- Tarragon
- Oregano
- Rosemary

Spices and condiments
- White wine
- Juniper berry
- Caraway
- Ginger
- Chili

Meats and dairy
- Speck

Cereals, pulses and nuts
- Wild black rice
- White beans
- Lentils

Salt-crusted sturgeon with fregola sarda risotto

Type
main course
Yield
4 servings
Preparation time
15 minutes
Cooking time
40 minutes

This dish showcases the technique of baking fish in a salt crust to keep it moist and tender. The fillet is first wrapped in Swiss chard leaves to prevent it from absorbing too much salt during the cooking process and is then served with a variation of traditional risotto made with fregola sarda, a pasta from Sardinia in Italy.

Utensils needed
- Saucepan
- Fine sieve
- Scissors
- Fishbone tweezers
- Baking tray
- Greaseproof paper
- Cutting board
- Non-stick pan
- Fine grater

Ingredients

Aromatic stock
- 1 litre water
- 3 thyme stems
- 2 rosemary stems
- 2 bay leaves
- 2 garlic cloves

Salt-crusted sturgeon
- 1 sturgeon fillet
- 1 kg coarse salt
- 6 Swiss chard leaves
- Salt and pepper

Fregola sarda risotto
- 1 white onion
- 6 Swiss chard stems
- 400 g fregola sarda
- 1 glass of white wine
- 2 tbsp unsalted butter
- 80 g grated pecorino
- Salt and pepper

Garnish
- 1 grapefruit
- 1 spring onion
- 80 g grated pecorino

Preparation

Aromatic stock
In a saucepan, simmer water with thyme, rosemary, bay leaves and garlic for 10 minutes. Filter the stock through a fine sieve.

Salt-crusted sturgeon
Rinse the sturgeon fillet under cold water and remove the skin and the fish bones. Season with salt and pepper. Separate the Swiss chard leaves from the stems and wrap the leaves around the fish fillet. On a baking tray lined with greaseproof paper, cover the sturgeon fillet with coarse salt. Bake in an oven preheated to 180 °C for 20 minutes.

Fregola sarda risotto
While the fish is baking, peel and finely chop the onion and finely dice the Swiss chard stems. In a pot, sauté the onions and the Swiss chard stems in olive oil. Add the fregola sarda and deglaze with white wine. Continuously add stock until the fregola sarda is cooked through. Stir in the butter and the grated pecorino and season with salt and pepper.

To serve
Peel and finely dice the grapefruit and finely chop the spring onion. Carefully remove the salt crust and slice the fillet. Serve alongside the fregola sarda risotto topped with the diced grapefruit, spring onions and grated pecorino.

Nutrition facts

	Per 100 g	Per recipe
Energy	311 kJ/74.3 kcal	12 954 kJ/3 096 kcal
Protein	4.4 g	182.0 g
Carbohydrate	7.1 g	295.0 g
Fibre	0.8 g	33.9 g
Sugar	0.6 g	25.5 g
Fat	2.8 g	115.0 g
Saturated fat	1.2 g	51.4 g
Sodium	139 mg	5 774 mg

Alvarez, M.R., Friedl, F.E., Johnson, J.S. & Hinsch, G.W. 1989. Factors affecting in vitro phagocytosis by oyster hemocytes. *Journal of Invertebrate Pathology*, 54(2): 233–241.

Bardot-Cambot, A. & Forest, V. 2013. Ostréiculture et mytiliculture à l'époque romaine? Des définitions modernes à l'épreuve de l'archéologie. *Revue archéologique*, 2: 367–388.

Bauchot, M.L. 1987. Poissons osseux. In: W. Fischer, M.L. Bauchot and M. Schneider, eds. *Méditerranée et mer Noire. Révision 1. Zone de pêche 37. Vol. 2. Fiches FAO d'identification pour les besoins de la pêche*, pp. 891–1421. Brussels, Commission des Communautés Européennes and Rome, FAO.

Botta, R., Asche, F., Borsum, J.S., & Camp, E.V. 2020. A review of global oyster aquaculture production and consumption. *Marine Policy*, 117: 103952.

Boudouresque, C.F. & Verlaque, M. 2007. Ecology of *Paracentrotus lividus*. In: J.M. Lawrence, ed. *Edible Sea Urchins: Biology and Ecology*, pp. 243–285. Vol. 37. Developments in aquaculture and fisheries science. Amsterdam, Elsevier.

Brundu, G., Farina, S., Guala, I., Guerzoni, S. & Pinna, S. 2020. Riccio di mare: ricerca e gestione della risorsa. *Il Pesche*, 20(2): 102–109.

Buschmann, A.H., Correa, J.A., Westermeier, R., Hernandez-Gonzalez, M.C. & Norambuena, R. 2001. Red algal farming in Chile: a review. *Aquaculture*, 194: 203–220.

Capillo, G., Sanfilippo, M., Aliko, V., Spano, A., Spinelli, A. & Manganaro, A. 2017. *Gracilaria gracilis*, Source of Agar: A Short Review. *Current Organic Chemistry*, 21(5): 380–386.

Chebanov, M.S. & Galich, E.V. 2013. Sturgeon hatchery manual. FAO Fisheries and Aquaculture Technical Paper No. 558. Rome, FAO. fao.org/3/i2144e/i2144e.pdf

Chopin, T. 2014. Seaweeds: Top mariculture crop, ecosystem service provider. *Global Aquaculture Advocate*, 17(5): 54–56.

Çiftci, Y., Üstündağ, C., Erteken, A., Özongun, M., Ceylan, B., Haşimoğlu A., Güneş, E., Yoseda, K., Sakamoto, F., Nezaki, G. & Hara, S. 2002. *Manual for the Seed Production of Turbot,* Psetta maxima *in the Black Sea*. Special Publication No. 2. Trabzon, Türkiye, Central Fisheries Research Institute, Ministry of Agriculture and Rural Affairs and Tokyo, Japan International Cooperation Agency.

CITES. 2023. Sturgeons. In: *CITES*. Geneva. Cited 22 January 2023. cites.org/eng/prog/sturgeon.php

Copp, G.H., Bianco, P.G., Bogutskaya, N.G., Eros, T., Falkal, I., Ferreira, M.T. *et al*. 2005. To be, or not to be, a non-native freshwater fish? *Journal of Applied Ichthyology*, 21: 242–262.

Craik, J.C.A. 1985. Egg quality and egg pigment content in salmonid fishes. *Aquaculture*, 47(1): 61–88.

EUMOFA. 2021. *The caviar market: production, trade, and consumption in and outside the EU an update of the 2018 report*. Brussels.

EUMOFA. 2022. *Oysters in the EU: Price structure in the supply chain focus on France, Ireland and the Netherlands*. Case study. Luxembourg, Publications Office of the European Union.

Eurostat. 2022. Production from aquaculture excluding hatcheries and nurseries (from 2008 onwards). In: *Eurostat*. Cited 6 December 2022. ec.europa.eu/eurostat/databrowser/view/FISH_AQ2A__custom_4058947/default/table?lang=en

FAO. 2019. *Regional Conference on river habitat restoration for inland fisheries in the Danube river basin and adjacent Black Sea areas. Conference Proceedings, 13–15 November 2018, Bucharest, Romania*. FAO Fisheries and Aquaculture Proceedings No. 63. Rome. https://doi.org/10.4060/ca5741en

FAO. 2022a. *Cyprinus carpio*. Cultured Aquatic Species Information Programme. In: *Fisheries and Aquaculture Division*. Rome. Cited 5 July 2022. fao.org/fishery/en/culturedspecies/cyprinus_carpio/en

FAO. 2022b. Common carp – Growth. In: *Aquaculture Feed and Fertilizer Resources Information System*. Rome. Cited 5 July 2022. https://www.fao.org/fishery/affris/species-profiles/common-carp/growth/en/

FAO. 2022c. *Mytilus galloprovincialis.* Cultured Aquatic Species Information Programme. In: *Fisheries and Aquaculture Division.* Rome. Cited 28 July 2022. fao.org/fishery/en/culturedspecies/mytilus_galloprovincialis/en

FAO. 2022d. *Sparus aurata.* Cultured Aquatic Species Information Programme. In: *Fisheries and Aquaculture Division.* Rome. Cited 31 July 2022. fao.org/fishery/en/culturedspecies/sparus_aurata/en

FAO. 2022e. *Mugil cephalus.* Cultured Aquatic Species Information Programme. In: *Fisheries and Aquaculture Division.* Rome. Cited 5 July 2022. fao.org/fishery/en/culturedspecies/mugil_cephalus/en

FAO. 2022f. *Oreochromis niloticus.* Cultured Aquatic Species Information Programme. In: *Fisheries and Aquaculture Division.* Rome. Cited 31 July 2022. fao.org/fishery/en/culturedspecies/oreochromis_niloticus/en

FAO. 2022g. *Scopthalmus maximus.* Cultured Aquatic Species Information Programme. In: *Fisheries and Aquaculture Division.* Rome. Cited 26 July 2022. fao.org/fishery/en/culturedspecies/Psetta_maxima/en

FAO. 2022h. *Dicentrarchus labrax.* Cultured Aquatic Species Information Programme. In: *Fisheries and Aquaculture Division.* Rome. Cited 31 July 2022. fao.org/fishery/en/culturedspecies/dicentrarchus_labrax/en

FAO. 2022i. *Onchorhynchus mykiss.* Cultured Aquatic Species Information Programme. In: *Fisheries and Aquaculture Division.* Rome. Cited 2 August 2022. fao.org/fishery/en/culturedspecies/oncorhynchus_mykiss/en

FAO. 2022j. *Magallana gigas.* Cultured Aquatic Species Information Programme. In: *Fisheries and Aquaculture Division.* Rome. Cited 29 July 2022. fao.org/fishery/en/culturedspecies/crassostrea_gigas_

FAO. 2023. Global aquaculture production quantity (1950–2020). In: *Fisheries and Aquaculture Division.* Rome. Cited 26 July 2022. fao.org/fishery/statistics-query/en/aquaculture/aquaculture_quantity

Figueras, A.J. 1989. Mussel culture in Spain and in France. *World Aquaculture,* 20(4): 8–17.

GFCM. 2021. *Report of the webinar on the status and future of seaweed farming in the Mediterranean and the Black Sea, Online, 15 July 2021.* Rome. fao.org/gfcm/technical-meetings/detail/en/c/1442598/

Günther, R. T. 1897. The oyster culture of the ancient Romans. *Journal of the Marine Biological Association of the United Kingdom,* 4(4): 360–365.

Gupta, M.V. & Belen O.A. 2004. A review of global tilapia farming practices. *Aquaculture Asia,* 9(1): 7–12.

Hanisak, M.D. & Ryther, J.H. 1984. Cultivation biology of *Gracilaria tikyahiae* in the USA. *Hydrobiologia,* 117: 295–298.

Harrison, I.J. 2002. Order Mugiliformes: Mugilidae. In: K.E. Carpenter, ed. *The living marine resources of the Western Central Atlantic. Volume 2: Bony fishes part 1 (Acipenseridae to Grammatidae),* pp. 1071–1085. FAO Species Identification Guide for Fishery Purposes and American Society of Ichthyologists and Herpetologists Special Publication No. 5. Rome, FAO. fao.org/3/y4161e/y4161e00.htm

Liao, Y.-C., Chang, C.-C., Nagarajan, D., Chen, C.-Y. & Chang, J.-S. 2021. Algae-derived hydrocolloids in foods: applications and health-related issues. *Bioengineered,* 12(1):3787–3801.

Liu, H. & Chang, Y-Q. 2015. Sea urchin aquaculture in China. In: N.P. Brown & S.D. Eddy, eds. *Echinoderm aquaculture,* pp. 127–146. Hoboken, New Jersey, John Wiley & Sons, Inc.

Manjappa, K., Keshavanath, P. & Gangadhara, B. 2011. Influence of sardine oil supplemented fish meal free diets on common carp (*Cyprinus carpio*) growth, carcass composition and digestive enzyme activity. *Journal of Fisheries and Aquatic Science,* 6(6): 604. https://dx.doi.org/10.3923/jfas.2011.604.613

Massa, F., Aydin, İ., Fezzardi, D., Akbulut, B., Atanasoff, A., Beken, A., & Bekh, V. 2021. Black Sea Aquaculture: Legacy, Challenges & Future Opportunities. *Aquaculture Studies,* 21: 181–220.

McBride, S. 2005. Sea Urchin Aquaculture. *American Fisheries Symposium,* 46: 179–208.

McHugh, D.J. 2003. *A guide to the seaweed industry.* FAO Fisheries Technical Paper No. 441. Rome, FAO. fao.org/3/y4765e/y4765e00.htm

Moreira, A., Cruz, S., Marques, R. & Cartaxana, P. 2021. The unexplored potential of green macroalgae in aquaculture. *Reviews in Aquaculture,* 14(1): 5–26.

Neori, A., Troell, M., Chopin, T., Yarish, C., Critchley, A. & Buschmann, A.H. 2007. The need for a balanced ecosystem approach to blue revolution aquaculture. *Environment: Science and Policy for Sustainable Development,* 49(3): 36–43.

Pander, J., Schnell, J., Sternecker, K. & Geist, J. 2009. The 'egg sandwich': a method for linking spatially resolved salmonid hatching rates with habitat variables in stream ecosystems. *Journal of Fish Biology,* 74, 683–690.

Pavlidis M. & Mylonas C.C. eds. 2011. *Sparidae: Biology and Aquaculture of Gilthead Sea Bream and Other Species.* Hoboken, USA, Wiley-Blackwell.

Saleh, M. 2008. Capture-based aquaculture of mullets in Egypt. In: A. Lovatelli & P.F. Holthus, eds., *Capture-based aquaculture. Global review,* pp. 109–126. FAO Fisheries Technical Paper No. 508. Rome, FAO. fao.org/3/i0254e/i0254e04.pdf

Saleh, M.A. & Salem, A.M. 2005. *National Aquaculture Sector Overview. Egypt.* FAO Inland Water Resources and Aquaculture Service (FIRI). Rome, FAO.

Seginer, I. 2016. Growth models of gilthead sea bream (*Sparus aurata* L.) for aquaculture: A review. *Aquacultural Engineering*, 70: 15–32.

Shelton, W.L. 2002. Tilapia culture in the 21st century. In: Guerrero, R.D. III and M.R. Guerrero-del Castillo eds., *Proceedings of the International Forum on Tilapia Farming in the 21st Century (Tilapia Forum 2002)*, pp. 1–20. Los Baños, Philippines, Philippine Fisheries Association.

Smith, C.L. 1990. Moronidae. In: J.C. Quero, J.C. Hureau, C. Karrer, A. Post & L. Saldanha, eds. *Check-list of the fishes of the eastern tropical Atlantic (CLOFETA) Vol. 2*, pp. 692–694. Lisbon, JNICT, Paris, SEI and Paris, UNESCO.

Stickney, R.R., ed. 2000. *Encyclopedia of aquaculture.* Hoboken, USA, Wiley.

Turolla, E. 2016. *Arcidae, Glycymerididae e Mytilidae.* Vol 2. Gasteropodi e bivalve marini dei mercati Europei. Ferrara, Italy, Istituto Delta Ecologia Applicata.

Turolla, E. 2020. *Pectinidae e Ostreidae.* Vol 3. Gasteropodi e bivalvi dei mercati europei. Ferrara, Italy, Istituto Delta Ecologia Applicata.

Verlaque, M. & Nedelec, H. 1983. Biologie de *Paracentrotus lividus* (Lamarck) sur substrat rocheux en Corse (Méditerranée, France): alimentation des adultes. *Vie et Milieu*, 33: 191–201.

Wang, M. & Lu, M. 2015. Tilapia polyculture: a global review. *Aquaculture Research*, 2015, 1–12.

Woynarovich, A., Hoitsy, G. & Moth-Poulsen, T. 2011. *Small-scale rainbow trout farming.* FAO Fisheries and Aquaculture Technical Paper No. 561. Rome, FAO. fao.org/3/i2125e/i2125e00.pdf

WWF. 2020. *The biology of Danube sturgeons.* Factsheet. Washington, DC.